Python, Deep Learning, and LLMs

Python, Deep Learning, and LLMs:
A Crash Course for Complete Beginners

Yegor Tkachenko, PhD

New York, New York

First Kindle Direct Publishing edition, 2025

Published in the United States of America by Yegor Tkachenko.

Library of Congress Cataloging-in-Publication data

Tkachenko, Yegor, author.
Python, Deep Learning, and LLMs: A Crash Course for Complete Beginners / Yegor Tkachenko.
Independently published via Kindle Direct Publishing, 2025.
Includes bibliographical references and index.
Identifiers: LCCN 2025918288 | ISBN 9781733902205 (paperback) | ISBN 9781733902229 (ebook)
LC record available at https://lccn.loc.gov/2025918288

While every effort has been made to ensure the accuracy and completeness of this book, it is provided "as is" without any warranty, express or implied. The author and publisher shall not be liable for any loss or damage caused, directly or indirectly, by the contents of this book.

The author and publisher shall have no responsibility for the persistency or accuracy of URLs for external or third-party internet websites referred to in this publication. No guarantee is given that any content on such websites is, or will remain, accurate or appropriate. Any access to such websites is undertaken at the reader's own risk.

All trademarks mentioned in this book belong to their respective owners and are used for identification purposes only.

10 9 8 7 6 5 4 3 2 1

To my family.

Contents

Contents

Abstract

Python, Deep Learning, and LLMs: A Crash Course for Complete Beginners is a coding and machine learning bootcamp in textbook form. You will get hands-on experience with the Python programming language, essential math concepts, neural nets – and, by the end, you will have coded and trained a pocket-sized language model. A high school math background is all you need to get started; no prior programming experience is required. Put in the work, and soon you will know, in concrete detail, how to create a miniature mind inside a chunk of silicon using code. If that is not magic, what is?

Preface

Large language models (LLMs) represent one of humanity's most exciting steps towards general artificial intelligence (AI). To outsiders, they can seem intimidating and opaque. Yet the core ideas behind LLMs are surprisingly accessible to anyone familiar with fairly basic concepts: derivatives of a function, simple arithmetic operations with lists and tables of numbers (more generally, vectors and matrices), basic probability, and a bit of programming.

Unfortunately, most introductions to LLMs fall into one of two extremes:

1. They are too superficial to be useful.

2. They assume advanced knowledge of programming, linear algebra, and optimization, making them inaccessible to newcomers.

This book aims to bridge this gap. It offers a self-contained, rigorous, yet approachable introduction to LLMs for readers with minimal math background and no prior programming or AI experience.

> **i** Who this book is for
>
> Ideally, you remember some high school algebra, recognize a derivative, and are ready to get your hands dirty with code. No prior experience with Python, machine learning, or language modeling is assumed. Refreshers on key concepts are included along the way.

We start from the ground up: Python programming, essential concepts from linear algebra and probability, and an introduction to regression and optimization. With these building blocks, we construct deep neural networks and, eventually, learn how they power LLMs that can hold conversations and tackle complex intellectual tasks.

All algorithms and methods are implemented in the Python programming language – the de facto standard for machine learning. Complete, executable code is provided throughout. I encourage you to type the code out, explore what it does, experiment, and, when you encounter errors or confusion, consult Google or a capable LLM. Active engagement is the fastest path to deep understanding, and you might be surprised by how much you can learn in a short time.

Given the vast scope of the subject, I have prioritized clarity and conciseness over exhaustive coverage. For those who want to dive deeper, I include further learning resources and references at the end of each chapter.

This book is not an encyclopedia – it is a language-immersion experience and a survival guide, designed to maximize your learning per hour invested. It requires work and will challenge you, but by the final chapter you won't just understand how LLMs work – you will have built a basic one yourself in Python. The goal is to equip you to meet the new era of AI with confidence and clarity.

> **i** Website
>
> Supplementary material is available on the book's official website: https://python2llms.org.

Yegor Tkachenko
New York, NY
August 21, 2025

Acknowledgments

This book grew out of the *Python Programming for Data Science* intensive course that I developed and have taught to students at Columbia Business School over the years. Their feedback has helped shape the course – and, by extension, this book – for which I am deeply grateful.

The material in this book distills more than a dozen years of my academic experience with Python and machine learning at Columbia and Stanford universities. I owe much of my understanding to the educators I encountered there. In particular, I would like to acknowledge Asim Ansari, Stephen Boyd, David Donoho, Peter Haas, Mark Hansen, Kamel Jedidi, Andrej Karpathy, Mykel Kochenderfer, Jared Lander, Peter Orbanz, and Rachel Schutt, whose intuitive and high-impact teaching styles remain a source of inspiration.

I am also grateful to many who provided their reactions and impressions regarding this book. In particular, I want to thank Iryna Kryvda and Matthew Roach, who reviewed early drafts of this book and provided valuable feedback. Your insights have made this work significantly better.

The typesetting of the book was done with the help of Quarto (https://quarto.org/), which relies on Jupyter (https://jupyter.org/), Pandoc (https://pandoc.org/), and LaTeX (https://www.latex-project.org/). The book is set using the KOMA-Script book class *scrbook*. Figures were generated with the help of PGFPlots (https://ctan.org/pkg/pgfplots), matplotlib (https://matplotlib.org/), and svgwrite (https://pypi.org/project/svgwrite/). Cover design was done by the author; it uses Russo One and EB Garamond fonts from Google Fonts (https://fonts.google.com/). OpenAI language models, mainly GPT-4o, o3, and o4-mini, assisted with the book editing (https://openai.com/).

1 Python programming basics

This chapter gets you started with the basics of Python programming. You will learn how to install Python, run code, and work with core programming concepts: variables, data types, data structures, control flow (like if – elif – else statements and for loops), functions, and libraries. These basics are the essential foundation for the rest of the book – and, hopefully, will serve you well for the rest of your life.

1.1 Why programming?

I like to think of a computer as a genie, waiting to be unleashed. With just a few carefully chosen commands, you can summon an army of virtual workers to automate tasks that would take humans lifetimes. Computers do not need rest, food, or salaries. As long as you know the set of instructions the computer can understand – that is, a *programming language* – the sky is the limit. You could build a billion-user web app, or a chatbot that understands humans. In many ways, programming is the closest thing we have to real-world magic in making colossal things possible with minimal effort.

Programming languages are unlike a human language in that they are designed to be completely unambiguous and specific – you have to tell a computer exactly what you want it to do – otherwise, it will not know how to act. (Arguably, with the arrival of large language models (LLMs), human languages have themselves become a kind of a programming language that you can use to control a computer – but we will stick with the programming vs. human language distinction.)

Fortunately, programming languages are generally much simpler in their grammar and vocabulary than human languages and so are also easier to learn. While mastery of a programming language certainly takes time and experience, this chapter gives you enough of the essentials to follow the rest of the book – even if you have never written a line of code before.

> 💡 Type in the code you see
>
> Type in and run the computer code I demonstrate. Do not just read – do. I am a big believer in muscle memory when it comes to coding. Use online resources if things are unclear. If you are stuck or see an error, search on Google, StackOverflow, or ask an advanced LLM like one of OpenAI's ChatGPT models. Professional developers do this every day. Over time, you will not need to rely on these tools as much, but you will never stop using them entirely if you continue to program.

1.2 Why Python?

The specific programming language we will use is **Python** (https://www.python.org/), originally conceived by Guido van Rossum and first published in 1991 [Wikc]. It is currently the go-to language for data science. It is simple, human-readable, and supported by an ecosystem of libraries for data processing, visualization, web interaction, and advanced math. It is open-source and free to use any way you like. In 2024, Python overtook JavaScript and became *the most popular* programming language on GitHub, one of the leading platforms for collaborative coding [Git24]. This is quite a symbolic milestone, given that JavaScript is the only programming language natively supported by all major browsers – and so is very widely used in web development and beyond – indeed, a worthy competitor to overcome in popularity.

Python is considered a *high-level* programming language, meaning its syntax is closer to natural human language – like English – than to low-level machine code, which consists of binary or Assembly instructions that communicate directly with a computer's hardware and are harder for a human to interpret. To illustrate the relative simplicity of Python syntax, Listing 1.1 shows code to print "Hello world" for Python and lower-level C++ and x86 Assembly.

This human-friendly design makes Python quick and easy to write, ideal for rapid development and prototyping. The trade-off, however, is that Python typically runs slower than lower-level languages like C, C++, or Assembly, which operate closer to the hardware and can generate more optimized machine instructions. Nevertheless, in many modern contexts – especially in research, data analysis, and prototyping – the cost of developer time often outweighs the benefits of marginal gains in code execution speed, making Python a highly practical choice.

Listing 1.1 Printing "Hello world" in Python, C++, and x86 Assembly

Python

```python
print("Hello, world!")
```

C++

```cpp
#include <iostream>

int main() {
    std::cout << "Hello, world!" << std::endl;
    return 0;
}
```

x86 Assembly (NASM syntax, Linux 64-bit)

```nasm
section .data
    msg db "Hello, world!", 0xA   ; store message + newline (0xA) in memory
    len equ $ - msg    ; get message length (from msg to current location $)

section .text
    global _start      ; declare the starting point of the program

_start:
    mov rax, 1         ; syscall: write
    mov rdi, 1         ; file descriptor: standard output (stdout)
    mov rsi, msg       ; message address
    mov rdx, len       ; message length
    syscall            ; make the system call to write the message to stdout

    mov rax, 60        ; syscall: exit
    xor rdi, rdi       ; set exit status to 0
    syscall            ; make the system call to exit
```

If speed becomes a concern, Python allows you to plug in optimized code from faster languages. In fact, many third-party libraries (code packages) that we will use from inside Python are actually written in one of other faster programming languages.

There have been efforts to develop a single language that combines the speed of a low-level language and the simplicity of a high-level language. Julia (https://julialang.org/), for example, is a strong contender in this space. However, Python's massive community, deep learning ecosystem, and overwhelming adoption by the machine learning industry and academia make it the clear choice for our purposes.

In some scenarios, using other programming languages can be unavoidable. For example, JavaScript – and, increasingly, TypeScript – is the standard for programming in web browsers. Swift is the primary language for building iOS apps, though alternatives like Objective-C or cross-platform frameworks (e.g., React Native or Flutter) are also used. In systems programming, languages like C, C++, and Rust are common due to their low-level control and performance.

That said, most core programming concepts – such as variables, control flow, functions, and data structures – are shared across languages. Once you have learned Python and become comfortable with these fundamentals, transitioning to another language is usually straightforward. With a bit of practice and reference to documentation, you can often get productive in a new language within a day or two.

1.3 Python installation

Install Anaconda distribution of Python (https://www.anaconda.com/download/success), which is free, works cross-platform, and includes a package manager `conda`.

After installing, launch your command line interface (CLI):

- **Mac**: Open `Terminal` (search for it via magnifying glass at top right corner of a Mac screen).
- **Windows**: Open `Anaconda PowerShell Prompt` (type it in the Start menu).

Upon CLI launch, you should see a console window pop up, awaiting your text input (Figure 1.1).

Verify Python installation by typing in the following commands into the CLI window, line by line, and hitting `enter` (`return`) after each line:

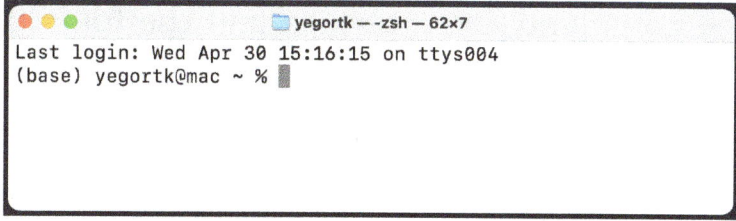

Figure 1.1: Command line interface upon launch (Mac version, Windows one looks similar)

```
python --version
pip --version
```

Commands you type will appear at the cursor rectangle in your CLI, after the percent sign that indicates input start (Figure 1.2). (`pip` is a package manager for installing libraries.) After running (executing) the two commands, you should see version numbers (Figure 1.2).

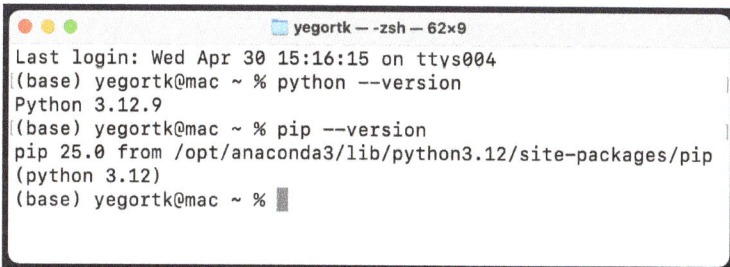

Figure 1.2: Command line interface – after command input

Note that these commands run in the CLI software – not within Python, which is a separate piece of software (we will see it a bit later).

Also consider installing Sublime Text 4 (https://www.sublimetext.com/download) – a minimal text editor that works well with raw Python code files.

1.4 The command line interface and the folder tree

The command line interface (CLI) is a text-based user interface for interacting with a computer. Programmers often use the CLI to navigate folders and run commands. While you are used to graphical interfaces, remote computers – such as web servers – are typically accessed and managed through the command line. The CLI window we opened (Figure 1.1) is commonly called the *terminal*. A *shell* is a program that runs inside the terminal and interprets your commands, for example, *bash (Bourne Again SHell)* – common on Linux / Mac, *zsh* – default shell on Mac, or *Anaconda PowerShell* for Windows.

Folders (directories) in an operating system are organized in a hierarchy, forming a tree, folders nesting within other folders. Any folder can contain files and other folders. The `root` folder is the topmost folder in the hierarchy. It contains all other folders and files directly or indirectly, but is not itself contained within any other folder.

Whenever we launch a CLI, we are placed in a specific folder along the folder tree. In general, any folder we are located at within CLI is called a *working directory*. The initial folder we get placed in upon CLI launch is sometimes called *home directory*.

The CLI provides multiple essential commands to help you navigate the folder tree:

- Use `cd` to change directories, `ls` to list files, `mkdir` to create a folder, `rm` or `del` to remove files/folders.
- Learn basic navigation: `pwd` prints current directory *path* (address), `cd ..` goes one level up the folder tree from current directory, `cd /` goes to the root folder, `cd ~` goes to home directory (just `cd` on Mac does the same), `cd <folderpath>` goes to specified folder, where `<folderpath>` string should be replaced with the address or path of the desired folder. Use quotes if the path contains spaces.
- Anything after `#` hash symbol is a comment and will not be executed.
- Some CLI commands differ between Mac (bash) and Windows (Anaconda PowerShell). There can be even more differences in case of other operating systems or other CLI software.

As an example, create a workspace folder using CLI commands:

```
# 1. Check path of current working directory
pwd


# 2. Go to Desktop folder
cd ~/Desktop
# tilde ~ indicates your home folder, which is,
# on MacOS and many Windows systems,
# the direct parent folder of the Desktop folder

# or, if you are already in your home folder, just do:
cd Desktop
# home folder is the default working directory
# upon opening MacOS Terminal and many Windows shells

# on Windows systems with OneDrive,
# you may need to use a different path:
# cd ~/OneDrive/Desktop


# 3. Make a folder python_book, and then a folder chapter_1
↪  within it
mkdir python_book
mkdir python_book/chapter_1


# 4. Move working directory to chapter_1 folder
cd "./python_book/chapter_1"
# . indicates current working folder
# . can usually be omitted at the beginning of nested paths
# but can be required, e.g., if a folder name matches a
↪  system command
# quotation marks are optional here,
# but are required if path contains a space
# cd python_book/chapter_1 would also generally work


# 5. Check current path again
pwd
```

Now, let us create a new file in the current working directory:

1 Python programming basics

```
# 1. Create file

# Mac
touch code_example.py

# Windows (Anaconda PowerShell)
ni code_example.py -type file

# 2. List files to confirm creation
ls

# 3. Move two levels up the folder tree - i.e., back to
↪  Desktop
cd ../..
```

Figure 1.3 shows how all these commands look executed on my Mac.

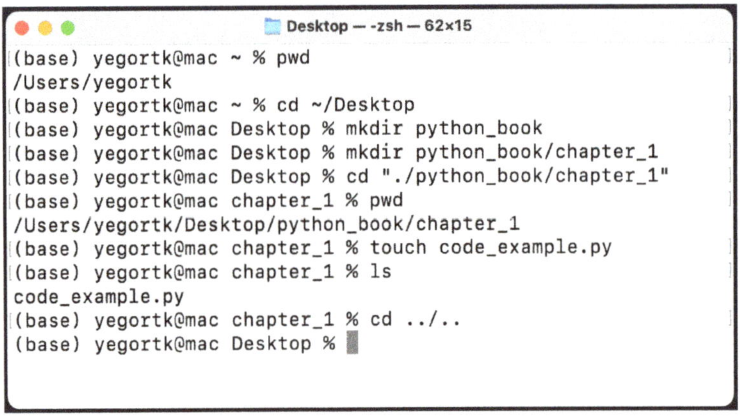

```
● ● ●                    📁 Desktop — -zsh — 62×15
(base) yegortk@mac ~ % pwd
/Users/yegortk
(base) yegortk@mac ~ % cd ~/Desktop
(base) yegortk@mac Desktop % mkdir python_book
(base) yegortk@mac Desktop % mkdir python_book/chapter_1
(base) yegortk@mac Desktop % cd "./python_book/chapter_1"
(base) yegortk@mac chapter_1 % pwd
/Users/yegortk/Desktop/python_book/chapter_1
(base) yegortk@mac chapter_1 % touch code_example.py
(base) yegortk@mac chapter_1 % ls
code_example.py
(base) yegortk@mac chapter_1 % cd ../..
(base) yegortk@mac Desktop % █
```

Figure 1.3: Command line interface – navigation

Comments:

- More path examples:

    - /Users is a path to a folder Users presumably located in the
      root folder.

- `../hello` is a path to a folder `hello` presumably contained in the parent folder of the current folder (i.e., both current folder and folder `hello` should be located in the same parent folder).

- An **absolute path** specifies a location in the file system independently of the current working directory. It starts with `/` (the root directory). For example, `/usr/local/bin` is an absolute path. The `~` symbol is a shell shorthand for the current user's home directory, and it expands to an absolute path. A **relative path**, on the other hand, describes a location in relation to the current working directory – it may start with `.` (current directory), `..` (parent directory), or directly with a folder or file name – in which case CLI looks for that folder or file name in the current working directory.

- Default path syntax differs between Mac and Windows by whether a forward slash (`/`) or a backslash (`\`) is used (`./Desktop/Folder` on Mac vs. `.\Desktop\Folder` on Windows), but Anaconda PowerShell CLI on Windows understands Mac-style paths, so we stick to those.

- **Useful trick:** If you drag and drop a folder or file into the CLI, its full path will automatically be pasted into the command line – this is convenient for quick `cd` navigation. Alternatively, you can right-click the folder or file to view and copy its path.

  - **Mac:** Right-click and choose `Get Info` , then copy the location shown under the `Where` section.
  - **Windows:** Right-click the item while holding `Shift` , then select `Copy as path` to copy the full file path to your clipboard.

- Desktop is just another folder – but it is rendered graphically as your home screen background.

- To access remote computers / servers, developers commonly use a tool called `ssh` (Secure Shell) from the CLI. This allows you to securely log into a remote machine using a command like `ssh username@server-address` . Once connected, you can navigate the file system, edit code, and run programs – all through the command line, just as if you were sitting in front of that machine.

- A shell is a very powerful piece of software with a lot of extra functionality and concepts that we won't cover here in detail. For example:

  - Shell users have specific permissions that determine whether they can read, write, or execute files and directories. These permissions are set by the system and are based on the user's identity and group membership. Permissions can be viewed, modified, or overridden using commands such as `chmod` (change permissions), `chown` (change ownership), and `sudo` (run commands with elevated privileges). You can search for these commands online to learn how they work.

— Shell includes some pre-defined *environment variables* (data available to the shell and programs it runs) – you can view these via `env` command.

— There are some useful shell operators. One example is the pipe operator `|` , which enables one to chain together different commands, passing text output from one command as input to another. Consider the command `env | wc -l`. Here `env` lists current environment variables, each printed on a new line; `|` passes this text as input to `wc -l` command; in `wc -l` , `wc` stands for word count, and the option `-l` tells the program to count lines instead of words. So `env | wc -l` returns the number of environmental variables.

— It is possible to run and manage multiple shell sessions within a single terminal window using a terminal multiplexer program, such as `screen` or `tmux` . For example, after launching a new shell session with the `screen` command, you can detach from it (`Ctrl+A` then `D`) and resume it later (`screen -ls` to see session ids and `screen -r <id>` to resume). This is quite useful when working with remote computers – a detached session continues running even if you close the terminal or lose your connection (e.g., over SSH).

1.5 Three ways to run Python code

Now that we can navigate to a folder of our choice with text commands, how do we run Python code from within it?

1.5.1 1. Interactive Python in CLI

Type `python` in CLI and press `enter`. You have entered Python mode – notice how your input indicator has changed to `>>>` (Figure 1.4). Try `print("Hello")`. Try `3+4`. This mode is fine for running short bits of code but is inconvenient for larger code volumes. Exit Python environment back into the CLI environment with `exit()` command or `Ctrl+Z` key combo.

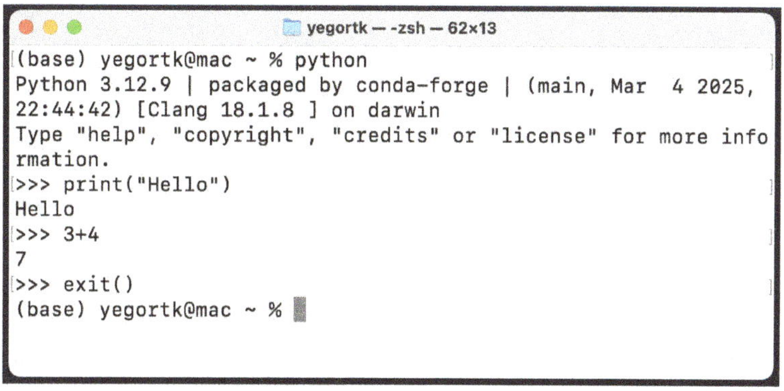

```
● ● ●                    yegortk — -zsh — 62×13
(base) yegortk@mac ~ % python
Python 3.12.9 | packaged by conda-forge | (main, Mar  4 2025,
22:44:42) [Clang 18.1.8 ] on darwin
Type "help", "copyright", "credits" or "license" for more info
rmation.
>>> print("Hello")
Hello
>>> 3+4
7
>>> exit()
(base) yegortk@mac ~ %
```

Figure 1.4: Python in command line interface

If you must use interactive interface like this for some reason, you may be better off using `ipython` instead of `python` – it comes preinstalled with conda – just type in `ipython` into shell instead of `python` and use the `%cpaste` magic command for pasting code, closing it with double dash `--` (see Figure 1.5). Command `%cpaste` in `ipython` ensures pasting large blocks of code with indentations does not result in malformed code – as it might when using the plain `python` interface.

1.5.2 2. Running code in .py files

Instead of entering Python code inside the console, we can first write it up in a text file and then tell Python to execute that file. Using Sublime Text 4, put `print("Hello friend")` in `code_example.py` file that we created earlier, then run command `python code_example.py` in CLI – inside the working directory where `code_example.py` file is located (Figure 1.6). This

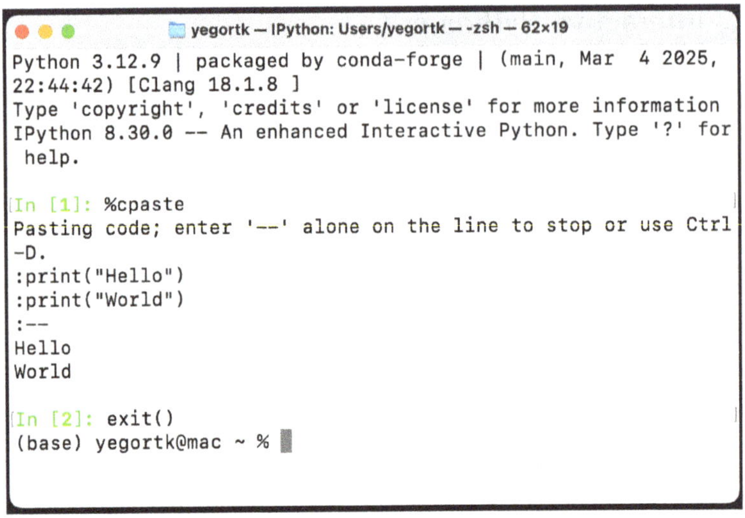

```
● ● ●              🗀 yegortk — IPython: Users/yegortk — -zsh — 62×19
Python 3.12.9 | packaged by conda-forge | (main, Mar  4 2025,
22:44:42) [Clang 18.1.8 ]
Type 'copyright', 'credits' or 'license' for more information
IPython 8.30.0 -- An enhanced Interactive Python. Type '?' for
 help.

In [1]: %cpaste
Pasting code; enter '--' alone on the line to stop or use Ctrl
-D.
:print("Hello")
:print("World")
:--
Hello
World

In [2]: exit()
(base) yegortk@mac ~ %
```

Figure 1.5: IPython in command line interface

executes your script. This mode of code execution is great for working with large code volumes as well as for scheduled code execution or for running code on servers.

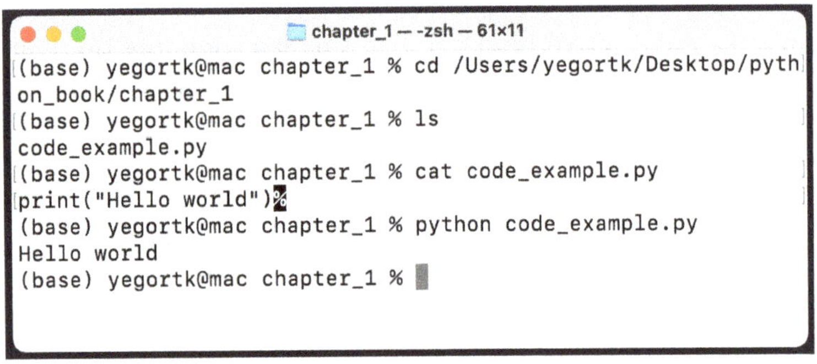

```
● ● ●                    🗀 chapter_1 — -zsh — 61×11
(base) yegortk@mac chapter_1 % cd /Users/yegortk/Desktop/pyth
on_book/chapter_1
(base) yegortk@mac chapter_1 % ls
code_example.py
(base) yegortk@mac chapter_1 % cat code_example.py
print("Hello world")
(base) yegortk@mac chapter_1 % python code_example.py
Hello world
(base) yegortk@mac chapter_1 %
```

Figure 1.6: Running .py file in command line interface

Command `cat` in Figure 1.6 lists the content of the file.

Note that a file with `.py` extension is just a regular text file. Python could as well execute a file without such an extension if it contained just the code. But `.py` extension does signal to the operating system – and to the casual observer – that one is dealing with Python code.

Sometimes you might see a "shebang" line at the top of python files: `#!/usr/bin/env python`. In Unix environments, this indicates an executable file that is meant to be interpreted and specifies the interpreter. It can be useful but is not necessary. For the purposes of this book, we will not use it. Read more on this here: [Ove].

1.5.3 3. Jupyter notebook

The last format for running Python code I will cover is called Jupyter notebook. With this approach, we use the CLI to launch a website, where we can type in code and execute it interactively within a very nice and convenient interface.

Run `jupyter notebook` command in CLI from inside your desired working directory. Once you execute this command, a Jupyter process will launch inside your CLI (Figure 1.7) and, simultaneously, a new window will pop up inside your browser (Figure 1.8).

Inside this browser window, click tab `New` in the upper right corner and select `Notebook`. This will launch yet another browser window that allows for code entry and creates a corresponding `.ipynb` file that will store the code and the output when you save it from inside the browser. You might get prompted to select the kernel – we will use Python 3 kernel (should be the default) – click `Select` on it.

The newly created notebook will be called `Untitled` (Figure 1.9) – you can rename it by clicking the name at the top.

This browser interface lets you write and run code in cells (by clicking play button in the top menu or using **Shift+Enter** combo to run code and go to next cell).

Jupyter notebooks are excellent for data exploration, visualization, and report generation. They store both code and its output and allow for rich formatting via Markdown markup language as well as for export of your work to different formats, such as raw `.py` Python files, HTML, PDF, etc.

For the rest of the book, you should be following the code and executing it via Jupyter notebook. You should think of the grayish cells within this

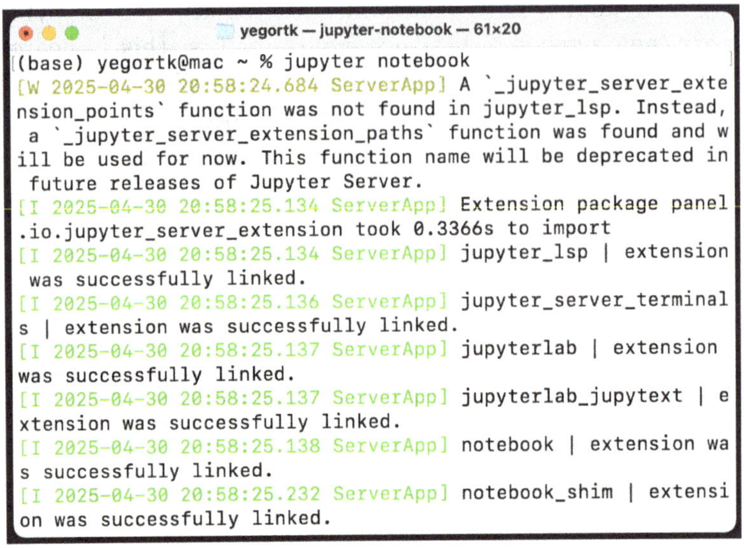

Figure 1.7: Jupyter notebook CLI launch

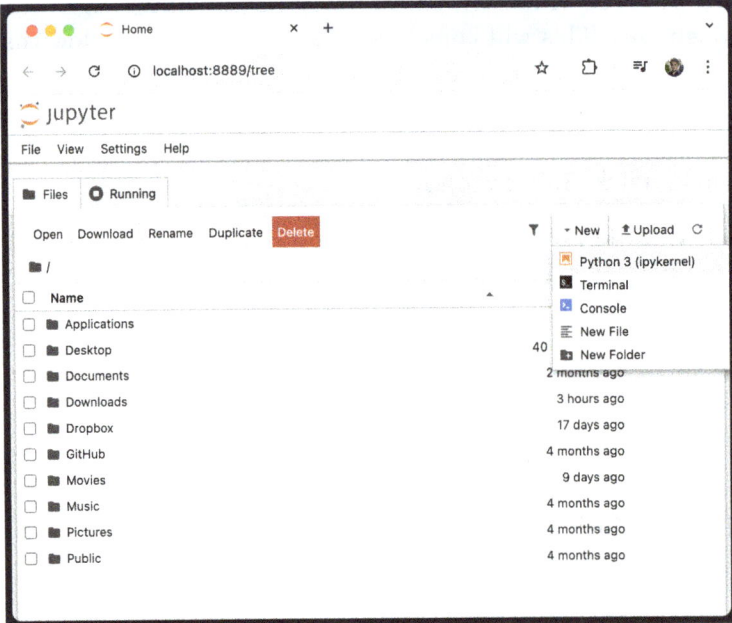

Figure 1.8: Jupyter notebook creation interface

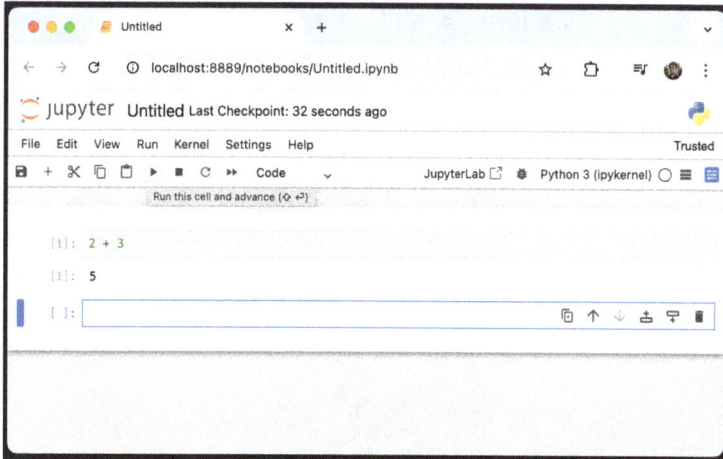

Figure 1.9: Jupyter notebook coding interface

book that contain Python code as corresponding to Jupyter notebook cells – the code output is presented below them.

1.6 The first taste of Python

We will proceed with `jupyter notebook` interface. (Launch it using instructions above.)

Please make sure to type and run the code that I present! If you encounter any errors – Google them or ask one of the popular LLMs.

Also note that Python code is platform-independent and can usually work without change across operating systems.

Python can be a calculator:

```
print(2+4)
print(2/5)
```

```
6
0.4
```

1 Python programming basics

Comments start with a hash `#` and are not executed:

```python
# This is a comment
2+2
```

```
4
```

Variables store data:

```python
a = 10
b = a + 5
print(b)
```

```
15
```

Variable names in Python can consist of English letters, numbers, and under-scores – and they *cannot start with a digit* (only a letter or an underscore). Variable names are *case-sensitive*, so `Data`, `data`, and `DATA` would all be distinct identifiers. Be careful when naming variables not to overwrite pre-existing objects in Python (e.g., function `print`).

> **i** Using `print()`
>
> Jupyter notebooks automatically display the output of the last line in a cell. However, this behavior is specific to the notebook interface – for instance, when running `.py` Python scripts, outputs are not shown unless explicitly printed. Even in Jupyter, only the last output in the cell is displayed by default, so if you want to show multiple outputs or ensure visibility across environments, it's a good practice to use `print()` for each value you want to display.

You can draw a random element from a list using a function `choice` from a library `random` like this:

```
# randomly selecting a travel destination
import random
cities = ["Paris", "Tokyo", "New York", "Sydney"]
city = random.choice(cities)
print(f"You are going to {city}!!")
```

```
You are going to Sydney!!
```

You can also prompt for input:

```
trip_type = input("Enter the type of trip (e.g., vacation,
↪  business, adventure): ")
print(f"You are going to {city} on {trip_type}.")
```

```
Enter the type of trip (e.g., vacation, business, adventure):
vacation
```

```
You are going to Sydney on vacation.
```

1.7 Data types: Integers, floats, booleans, strings

Data types refer to the kinds of values that *variables* can hold. They define what operations can be performed on the data and how it is stored in memory.

This is a variable storing an integer value (we *assign* an integer value to a variable named a):

```
a = 45
print(a)
```

```
45
```

To be more precise, variables contain an address pointing to a specific area in computer memory where the actual information is stored:

```
print(id(a)) # id of variable a and, usually, its memory
  ↳ address
print(hex(id(a))) # same in hexadecimal number system (base
  ↳ 16 vs. usual 10)
```

```
4343475944
0x102e42ee8
```

Integer value in binary:

```
print(a)
print(bin(a))
```

```
45
0b101101
```

> **i Bits**
>
> In a typical computer, all information stored in memory is ultimately represented as *bits* – binary digits (0 / 1) – which correspond to the physical on / off states of transistors in the hardware. Numbers, for instance, can be expressed in various bases but always reduce to a sequence of bits:
>
> - Decimal (base 10):
>
>   - Number 45 is written as
>
> $$45 = 4 * 10^1 + 5 * 10^0.$$
>
>   - In decimal, digits (here, 45) are multiplied by powers of 10 – thus the name *decimal*. (Recall that any non-zero number raised to the power of zero is equal to one.)

- Binary (base 2):

  - The same value appears as 101101 (also written as $0b101101$), since

  $$45 = 1 * 2^5 + 0 * 2^4 + 1 * 2^3 + 1 * 2^2 + 0 * 2^1 + 1 * 2^0.$$

  - In binary, digits (here, 101101) are multiplied by powers of 2. So when we write 45 in binary, we are asking what combination of powers of 2 adds up to 45.

- Hex / hexadecimal (base 16):

  - In hex, digits 0–9 are followed by a–f for values 10–15.
  - Hex number $2d$ (also written as $0x2d$) represents decimal number 45, since

  $$45 = 2 * 16^1 + 13 * 16^0$$

  (with $13 = d$).
  - In hex, digits (here, $2d$) are multiplied by powers of 16. When we write 45 in hex, we are asking what combination of powers of 16 adds up to 45.

Further, arithmetic operations can be constructed from logical operations like **and** and **or** applied to binary representations [Wikb]. You can read more about binary representation of numbers here: [Wika]. You can read about constructing a computer from scratch from binary numbers and logic operations here: [NS08].

1.7.1 Integers vs. floats

In Python, numbers come in different types, with the most common being `int` (integer) and `float` (floating-point number).

- `int` : Exact whole numbers, no decimal (floating) point, no rounding or precision error. For example, non-negative integers can be used for counting and representing discrete quantities.

- `float` : *Real* numbers [Wike] with a decimal point representation, often denoted by ℝ. These numbers are used to measure continuous quantities – we use them all the time for the everyday arithmetic. Internally, Python uses binary floating-point representation, which can introduce rounding errors for certain decimal values. The simplest example of this is the infinite decimal representation of the fraction `1/3` as `0.33333...` , which requires infinite number of digits for exactness – and, if represented by a finite binary number (without clever tricks), would involve cutting off the infinite expansion and rounding it. We will see more examples of this below.

```
print(type(2))      # int
print(type(2.0))    # float
```

```
<class 'int'>
<class 'float'>
```

Arithmetic operations:

- `+` , `-` , `*` , `/` for addition, subtraction, multiplication, and division respectively
- `**` for power
- `//` for floor division (throws away remainder)
- `%` for modulo (returns remainder)

Large integers are handled well by Python:

```
2**9999 # works (large int)
```

```
99753155844037919244187108134179254191174841594309622742600
4474926471941511097331595998084201809729894966556471160456213
5778245674706890558796892966048161978927865023396897263382
62327563302994776027504345909665577125430423030905234275453743
30448124440452449474190046269708166289253107841547369512784
561940326125483219372205233799358134927266114342690808471578
8
...
```

Large floats may *overflow* – the program cannot store the number in (capped) memory:

```
2.0**9999 # overflow
```

`OverflowError: (34, 'Result too large')`

Floats can also lead to slight inaccuracies because their binary representation in the computer is inherently approximate:

```
print(0.14 * 100)          # evaluates as not exactly 14.0!
print(round(0.14 * 100, 1))  # rounding helps
```

```
14.000000000000002
14.0
```

> **i** Built-in functions
>
> Rounding function `round()` is one of the built-in functions in Python – you can explore more of these here: [Doc].

Specialized libraries can help with arbitrary precision floats (e.g., see `decimal` or `mpmath`).

Converting types:

```
print(int(1.7))      # 1
print(float(2))      # 2.0
print(2 - 1.0)       # 1.0 – mixing integers and floats
 ↳   reduces to a float
```

```
1
2.0
1.0
```

1.7.2 Booleans

In Python, booleans represent truth values and are written as **True** (1), **False** (0) – note the capitalization. Comparison operations return booleans:

```python
print(3 < 2)     # False (less than)
print(3/2 >= 1)  # True (greater or equal to)
print(3/2 != 1)  # True (not equal)
print(3/2 == 1)  # False (equal)
```

```
False
True
True
False
```

Logical operations for combining boolean variables include **and**, **or**, **not** :

```python
print(True and False)  # False
print(True or False)   # True
print(not True)        # False
```

```
False
True
False
```

These are useful for building more complex conditions, like:

```python
x = 5
print(x > 0 and x < 10)  # True (x is between 0 and 10)
```

```
True
```

Python treats **True** as 1 and **False** as 0 in calculations:

```
print(1.0 * True)  # 1.0
print(1.0 * False)  # 0.0
print(True + False + True)  # 2
```

```
1.0
0.0
2
```

1.7.3 Strings

Text is stored in strings:

```
print("Hello")
print('He"l"lo') # single or double quotes outside are fine

print("yegor\ttkachenko") # special character \t - tab
print("yegor\ntkachenko") # special character \n - new line
```

```
Hello
He"l"lo
yegor    tkachenko
yegor
tkachenko
```

To treat backslashes literally, use raw strings:

```
print("C:\new_folder") # \n is interpreted as a newline
↪    character
```

```
C:
ew_folder
```

```
print(r"C:\new_folder") # prints string as is thanks to
 ↳  r"..."
```

```
C:\new_folder
```

Variables can be embedded in strings using formatted strings:

```
a = "Bob"
print("Hello {}".format(a))
print("Hello {a}") # regular string just prints curly
 ↳  brackets
print(f"Hello {a}") # f-string f"..." for embedding variables
print(f"Bob is a {1000000:,.2f}")
# two numbers after decimal point, comma thousands separator
```

```
Hello Bob
Hello {a}
Hello Bob
Bob is a 1,000,000.00
```

Strings come with functions that can modify them:

```
print("hello".upper())
print("hello".title())
print("Hello".lower())
print("  yegor".strip()) # remove trailing spaces
```

```
HELLO
Hello
hello
yegor
```

New strings can be constructed from strings and split into substrings:

```
print(" ".join(["You","are","great"]))
print("John " + "Doe")

print("hello"[1:3])
# indexing (zero-based, from el #1 including to el #3
↪  not-including)

print("yegor tkachenko".split()) # string splitting (on
↪  space, by default)
print((4 * "super! ").strip()) # repetition via
↪  multiplication
```

```
You are great
John Doe
el
['yegor', 'tkachenko']
super! super! super! super!
```

Convert strings and numbers:

```
print(str(10))
print(int("3"))
print(float("3.56"))
```

```
10
3
3.56
```

Lexicographic ordering:

```
print("a" < "b") # lexicographic ordering - used to sort text
print(sorted("bac"))
```

```
True
['a', 'b', 'c']
```

Common issues:

```
# COMMON ERROR!! instead of
# a = 'Alex'
a = print('Alex')
```

```
Alex
```

```
a # contains nothing (None) - because print() command returns
↳   nothing
```

```
type(a)
```

```
NoneType
```

Python only knows how to concatenate two strings, not, for example, a string and an integer:

```
"yegor is a " + 10
```

```
TypeError: can only concatenate str (not "int") to str
```

1.7.4 Scientific notation

You will occasionally encounter Python numerical output formatted using scientific notation with the e symbol. For example:

- 1.23e4 means 1.23×10^4 or 12300
- 5.6e-2 means 5.6×10^{-2} or 0.056

Example of formatting:

```
print("{:.2e}".format(12300)) # convert to sci notation
print("{:.2f}".format(1.23e4)) # convert from sci notation
```

```
1.23e+04
12300.00
```

1.7.5 Advanced: Binary data and encodings

We have already seen earlier that numbers can be encoded in a binary format – as a sequence of *bits* or 0 / 1 values. For example, an integer is encoded to binary like this:

```
bin(45)
```

```
'0b101101'
```

With some more effort, we can also encode to binary the real (floating point) numbers. Commonly used rules defining how to do this are specified in IEEE 754 standard – we won't cover these in detail here. Example:

```
import numpy as np  # using numpy library, aliased as np
np.binary_repr(np.float32(-45.32).view(np.int32), width=32)
```

```
'11000010001101010100011110101110'
```

However, it is not just numbers that we can do this to – all data on a computer ends up represented (encoded) in binary. How exactly data is represented in binary is dictated by a set of specific encoding rules.

Encoding is the process of converting human-readable information – such as text, images, or entire documents – into a format suitable for computer storage or transmission. *Decoding* reverses this process.

For example, *ASCII* (American Standard Code for Information Interchange) is a simple encoding / table that defines a correspondence between 128 commonly used English characters (including letters, punctuation, space, etc.) and 128 integer indices (0-127). Because $2^7 = 128$, 7 bits are sufficient to encode all ASCII characters. An example of ASCII encoding use:

```
# encode character to ASCII index
print(ord('A')) # integer
print(bin(ord('A'))) # binary
```

```
# decode ASCII code to character
print(chr(65)) # character
```

```
65
0b1000001
A
```

Historically, it is more common to use a sequence of eight bits (0 / 1 values) as a unit of digital information ($8 = 2^3$ and computer scientists like the powers of two). Such an 8-bit sequence is called a *byte*. A byte, comprised by eight bits, can represent $2^8 = 256$ different discrete elements.

Another noteworthy feature of a byte is that $256 = 16^2$ and so a byte can be exactly represented by a combination of two hexadecimal (base-16) digits that can each assume values 0-9 continuing through a-f (for values 10-15). Raw byte data is often presented using pairs of hex digits due to the compactness of such a format.

One byte is enough to encode all 128 ASCII characters – with some byte values left unused. There exist extended ASCII tables (such as Windows-1252) that incorporate additional characters – to the total of 256 – using the extra 8th bit in a byte.

The most widely used text encoding nowadays is UTF-8, which supports all Unicode characters (over 1 million) and represents them using 1 to 4 bytes (shorter representations for more frequently used characters and vice versa).

Python provides a string-like format `b"..."` to store raw byte data. This format is called a *bytes object* or a *b-string*. A byte object represents a sequence of bytes – it often arises when working with text, images, files, or network data. Encoding a string as a bytes object using UTF-8:

```
text = "hello"  # regular string
encoded = text.encode("utf-8")  # convert string to bytes
  ↳  object according to utf-8 symbol-byte correspondence
encoded  # still rendered as b-string for display, but
  ↳  actually has binary underneath
```

```
b'hello'
```

Underlying binary:

```
' '.join(format(byte, '08b') for byte in encoded)
```

```
'01101000 01100101 01101100 01101100 01101111'
```

Sequence of bytes as a sequence of their 0-255 integer values:

```
list(encoded)
```

```
[104, 101, 108, 108, 111]
```

Sequence of bytes as a sequence of hex digit pairs:

```
encoded.hex()
```

```
'68656c6c6f'
```

Bytes from hex:

```
bytes.fromhex(encoded.hex())
```

```
b'hello'
```

Decoding a regular string from bytes:

```
decoded = encoded.decode("utf-8")   # string from bytes
decoded
```

```
'hello'
```

Decoding binary data using the wrong encoding (e.g., decoding UTF-8 bytes as Latin-1) can lead to errors or corrupted output. See more on this topic here: [Sol].

1.8 Libraries

Libraries or *packages* are sets of Python code someone else wrote that you can import and use. Some come pre-installed with Anaconda Python distribution, others need to be installed from the Internet.

You can install new libraries, for example, with `pip`. To install *qrcode* package, you would run `pip install qrcode` from CLI (outside Python). Or, you could run `!pip install qrcode` inside Jupyter notebook, using `!` prefix to indicate you want the command to be run outside Python – in CLI. (`!` prefix works in Jupyter and IPython, but not in raw Python.)

```
!pip install qrcode
# the line after ! operator is run in CLI outside Python to
↪  install qrcode

import qrcode
img = qrcode.make("https://www.google.com")
img.save("qrcode_test.png")
```

Generated QR code (point your smartphone camera at it):

Figure 1.10: QR code

Package manager `conda` can also be used for library installation, as we will see later in the book.

---

i Virtual environments

By default, libraries installed with `pip` or `conda` are placed into your system or base Python environment. Over time, this can lead to clutter or version conflicts – situations where different projects require incompatible versions of the same library. To avoid these issues, you can use *virtual environments* – isolated environments that contain their own independent sets of libraries. Two common ways to create them are:

- `python -m venv <envname>` – a standard Python virtual environment [Pyt]

- `conda create -n <envname>` – a `conda`-managed virtual environment [Con]

(Replace `<envname>` with a preferred name.)
While we will not worry about this here, you can use the references above to learn about creating and using virtual environments.

1.9 Conditioning: if – elif – else

Conditional statements let your program make decisions based on whether something is true or false. They use `if`, `elif`, and `else` to decide what code to run.

Each block of code that depends on a condition is *indented* – usually by 4 spaces or a tab – to show it is a part of that condition.

```python
gmat = 710
if gmat > 700:
    print("admit")
else:
    print("reject")
```

```
admit
```

```
if gmat > 740:
    print('admit')
elif gmat > 720:
    print('waitlist')
else:
    print('reject')
```

```
reject
```

Use `and`, `or` for compound conditions.

We can also nest conditions within each other by specifying indentation appropriately:

```
a = 7
b = 10
c = 3

if (a > 5) or (c <= 1):
    if b > 9:
        print("Yay")
```

```
Yay
```

1.10 List and tuple data structures

Lists are ordered, mutable sequences:

```
a = [1,2,3]   # new list
print(a[0])   # zero-based indexing - first element

a.append(4)   # add new element at the end
print(a)
print(sum(a))
```

```
1
[1, 2, 3, 4]
10
```

Slicing and indexing:

```
print(a[1:3])  # second a[1] and third a[2] elements (last
↳  index is not-including)
print(a[-1])  # last element of a
print(len(a))  # list length
```

```
[2, 3]
4
4
```

```
a = a + [4, 6] # using + operator to concatenate two lists
```

```
a.extend([3,4,5]) # using extend command to do the same
↳  in-place (without assignment)
print(a)
```

```
[1, 2, 3, 4, 4, 6, 3, 4, 5]
```

Note:

- Python uses zero-based indexing, where first element in a list is indicated by 0 , second element – by 1 , and so on.
- When slicing is used, e.g., a[2:5] – this indicates elements index 2, 3, 4 will be included, but element with index 5 will not be included.
- a[2:] means all elements starting with index 2 until the end of the list will be included.
- Negative indexing means indexing from the back of the list: a[-1] gives the last element, a[-2] – second to last element, a[-2:] would give the last two elements in a list.

- `a[0:5]` and `a[:5]` give identical result.
- `a.append()` and `a.extend()` are a so-called *in-place* operations. This means list `a` is modified in memory – results of append or extend do not need to be assigned to variable `a` to be preserved / saved. In contrast, running `a + [4, 6]` does not change `a` – the output needs to be assigned to `a` to be preserved, as in `a = a + [4, 6]` – otherwise output is discarded.

Tuples are like lists but immutable (not changeable).

```
a = (1, 2, 3)    # round brackets! not square
a[2] = 1         # ERROR: Tuples are immutable - you cannot
 ↳  change their elements
```

```
TypeError: 'tuple' object does not support item assignment
```

```
list(a) # conversion to list
```

```
[1, 2, 3]
```

Tuples are often used instead of lists for data that is not expected to change – the tuple data structure prevents accidental addition and removal of elements.

1.11 Repetition: for loops and while statement

A *for loop* lets you repeat an action for each item in a sequence (like a list, a string, or a range of numbers). It is useful when you want to do something multiple times or process each item in a group.

```
for x in a:    # x sequentially assumes value of each element
 ↳  in a
    print(x**2)
```

```
1
4
9
```

`range()` function helps to generate number sequences of specified length:

```
for i in range(10):
    print(i)
```

```
0
1
2
3
4
5
6
7
8
9
```

List comprehension is a more compact way run a loop and store the output in a list:

```
[i**2 for i in range(10)]
```

```
[0, 1, 4, 9, 16, 25, 36, 49, 64, 81]
```

Statements **break** and **continue** can be used to control loops. Use of **continue** skips an iteration. Use of **break** stops the loop early.

```
# printing only even numbers
for i in range(10):
    if i % 2 == 1:  # if number i is odd
        continue  # skip to the next iteration of the loop
    print(i)
```

```
0
2
4
6
8
```

```
a = [3,4,5,2,5]
a
```

```
[3, 4, 5, 2, 5]
```

```
for i in range(len(a)):
    if a[i] == 5:
        print(i)
        break  # stop the for loop
```

```
2
```

Another way to do code repetition is to use **while** keyword. A while loop runs as long as a condition is true:

```
x = 0
while x < 10:
    x += 1
    print(x)
```

```
1
2
3
4
5
6
7
8
9
10
```

Notice that `while True:` loop would potentially run forever, until inter-rupted. To avoid the infinite run, you can use a `break` condition or other logic to eventually exit the loop – otherwise it will run forever and could freeze your program. That said, an infinite `while` loop is also often used intentionally – for example, to run processes that should do something at regular intervals forever – just like a web server that constantly listens for incoming connections or an operating system that keeps responding to user inputs or system events – both are essentially big `while True` loops (although not quite literally in the case of operating systems, which are a bit more complex).

1.12 Functions

Functions allow you to package logic into reusable blocks. Instead of writing the same code repeatedly, you define a function once and then call it whenever you need it.

You define a function using the `def` keyword:

```
def mean_value(a):    # a is a parameter that this function
↳  accepts
    return sum(a) / len(a)
```

The variable `a` above is a *parameter* of a function – it appears in the function definition and acts as a placeholder. When you call the function, the value you provide as input is an *argument*. (Despite this technical distinction, these terms are often used interchangeably in practice to refer to "what goes into the function.") You can now use this function like so:

```
mean_value([1,2,3])
```

```
2.0
```

This function takes a list of numbers and returns their mean (average). It's much cleaner than writing `sum(x) / len(x)` every time.

Functions can also have *default argument values*, which are used if the caller does not provide a value for that parameter:

```python
def take_to_power(x, p=1):  # two parameters, one with
 ↳   default value
    return x**p

print(take_to_power(2))      # uses default p=1 >> 2**1 = 2
print(take_to_power(2,2))    # overrides p >> 2**2 = 4
```

```
2
4
```

Lambda functions are an alternative to `def`, used for a single-line function definition:

```python
sq = lambda x: x**2
sq(10)
```

```
100
```

Lambdas are useful for short, throwaway functions, especially in the context of sorting and data transformation.

Note: Differences between return and print

Consider two functions that accept no arguments:

```python
def hello():
    return "hello"
```

```python
def hello_p():
    print("hello")
    # this function would print hello,
    # but it returns None variable - can cause bugs if you
     ↳   forget to return
    # having no return keyword is equivalent to indicating
     ↳   return None
```

Here:

- `hello()` returns the string "hello". You can store it in a variable, use it in expressions, etc.
- `hello_p()` prints to the screen but returns nothing (None). It is for side effects (like displaying a message), not for producing a reusable result.

```
a = hello()
type(a)
```

```
str
```

```
b = hello_p()
type(b)
```

```
hello
```

```
NoneType
```

1.13 Set and dictionary data structures

Sets are collections of unordered unique elements:

```
s = {1,2,3,3}
s
```

```
{1, 2, 3}
```

Converting a list to a set is a useful trick to count only the unique elements:

```
len(set([1,2,3,3]))
```

```
3
```

1 Python programming basics

Union and intersection of sets:

```
print({1,2} | {2,3})
print({1,2} & {2,3})
```

```
{1, 2, 3}
{2}
```

Dictionaries are sets of *keys*, where each key is paired with a value:

```
# new dict
microsoft = {
    "name": "Microsoft",
    "ticker": "MSFT",
    "price": 418
    }

print(microsoft)
print(microsoft["ticker"])  # MSFT - value stored in key
 ↳    `ticker`
```

```
{'name': 'Microsoft', 'ticker': 'MSFT', 'price': 418}
MSFT
```

```
microsoft["competitors"] = ["Google", "Apple"] # assign a
 ↳    value to a key
microsoft
```

```
{'name': 'Microsoft',
 'ticker': 'MSFT',
 'price': 418,
 'competitors': ['Google', 'Apple']}
```

```
print(microsoft.keys())  # all dict keys (useful for looping)

del microsoft['competitors'] # delete key + its value

print(microsoft)
```

```
dict_keys(['name', 'ticker', 'price', 'competitors'])
{'name': 'Microsoft', 'ticker': 'MSFT', 'price': 418}
```

> 💡 Table as a list of dictionaries
>
> A list of dictionaries can be used to represent a table-like structure,
> where each dictionary corresponds to a row, and the dictionary keys
> act as column names.

Simple tuples (but not lists) can be used as keys in the dictionary:

```
dbin = {
    (0,0) : 0,
    (0,1) : 1,
    (1,0) : 2,
    (1,1) : 3,
}
print(dbin[(0,1)])
```

```
1
```

1.14 Misc topics: Mutability and scope

1.14.1 Mutability

```
a = 10
b = a
```

```
a += 1 # a = a + 1

print(a)
print(b)
```

```
11
10
```

Integers are immutable in Python. When an integer is modified in Python, a new object is created elsewhere in memory (unless it already exists there). So when we increment `a += 1` , the number where `a` used to point does not change – instead, variable `a` starts pointing to a new place in memory.

```
a = [10, 20]
b = a
a.append(30)

print(a)
print(b)
```

```
[10, 20, 30]
[10, 20, 30]
```

Lists are mutable. Python puts a list somewhere in memory upon assignment, variable `a` stores address to that list, we then copy the address for the same list to variable `b` , so now both variables point to the same spot in memory. When we change the list by calling `a.append(30)` , the list that both variables point to changes – so both variables return the same changed list.

Use `copy` or `deepcopy` *if you want to edit a copy of a mutable structure without altering the original* (this can be useful, e.g., if you need to compare the new vs. the old versions of the structure).

```
from copy import deepcopy
b = deepcopy(a)
a.append(100)
print(a)
print(b)
```

```
[10, 20, 30, 100]
[10, 20, 30]
```

1.14.2 Scope

Scope refers to the part of your code where a variable is visible or accessible. In Python, variables created inside a function exist only within that function – they have local scope. Variables defined outside any function have global scope and are accessible throughout the file, unless overridden locally.

```
x = 300           # global variable
def test():
    x = 200       # local variable (only exists inside the
 ↪  function, locally overrides global x)
    print(x)
test()
print(x)          # refers to the global x
```

```
200
300
```

You could tell Python to access the global variable locally, but it is dangerous and is discouraged:

```
x = 300           # global variable
def test():
    global x      # tells Python to use global x
    x = 200       # changes global variable
    print(x)
```

```
test()
print(x)
```

```
200
200
```

> ⚠ Avoid `global` command
>
> Avoid using `global` unless absolutely necessary. It makes code harder to reason about and debug.

Prefer returning values from functions instead:

```
x = 300
def test():
    x = 200
    print(x)
    return x
x = test()
print(x)
```

```
200
200
```

1.15 Coding exercise

Let us use your new Python skills to explore a real book (in public domain) – "Alice's Adventures in Wonderland" by Lewis Carroll – and analyze its text.

1.15.1 Problem 1 — Letter frequency from a web page

Download the full text of *Alice in Wonderland* from Project Gutenberg site and report 10 most frequent letters in descending order. Treat upper and lower case letters as identical.

- URL: https://www.gutenberg.org/files/11/11-h/11-h.htm.

Solution:

Modern websites are written in HTML [Wikd] – hypertext markup language
– which browsers can understand and render for their users as a web page.
If you use Google Chrome, you can right-click on a web page and select
`Inspect` option to view the underlying HTML.

We will use Python to download HTML content from the provided URL,
extract raw text from it, clean it up, and perform the required character
count. First, let us download the data:

```
# !pip install beautifulsoup4

import requests   # library to download from a url
from bs4 import BeautifulSoup   # library to parse html and
↳   extract raw text

url = 'https://www.gutenberg.org/files/11/11-h/11-h.htm'

raw_bytes = requests.get(url).content # download raw bytes of
↳   html page
html = raw_bytes.decode('utf-8') # decode manually into text
↳   string

# alternatively, you could let requests auto-resolve the
↳   encoding
# html = requests.get(url).text


print(html[:500]) # print first 500 characters
```

```
<!DOCTYPE html PUBLIC "-//W3C//DTD XHTML 1.0 Strict//EN"
"http://www.w3.org/TR/xhtml1/DTD/xhtml1-strict.dtd">  <html
xmlns="http://www.w3.org/1999/xhtml" xml:lang="en"
lang="en">  <head>  <meta http-equiv="Content-Type"
content="text/html;charset=utf-8" />  <meta http-
equiv="Content-Style-Type" content="text/css" />
```

```
<title>Alice's Adventures in Wonderland | Project
Gutenberg</title>  <link rel="coverpage"
href="images/cover.jpg"/>  <style type="text/css">    body {
margin-left: 10%;        ma
```

We can save and read back the html file like this:

```
# save to html file
with open("alice.html", "w", encoding="utf-8") as f:
    f.write(html)

# read back into html variable
with open("alice.html", "r", encoding="utf-8") as f:
    html = f.read()
```

> **i** Context manager `with` for opening files
>
> The `with` construction above is a Python idiom for safely opening a file. `with` statement is called a context manager and it ensures that the file is automatically closed after the indented block is finished, even if an error occurs. `open("alice_in_wonderland.html", "r", encoding="utf-8")` opens the file named "alice_in_wonderland.html" in read mode (`r`) using UTF-8 encoding. `w` means write mode.
> For example, `with` read-mode code block above is equivalent to the following code, where `try` catches any error and allows code execution to continue, ensuring file connection is closed:
>
> ```
> f = open("alice_in_wonderland.html", "r",
>  ↪ encoding="utf-8")
> try:
>     html = f.read()
> finally:
>     f.close()
> ```

Now, we need to get raw text of the story out of HTML – we can use BeautifulSoup object for that:

```
soup = BeautifulSoup(html, 'html.parser')   # parse html
text = soup.get_text()   # raw text from html as a single
↳  string
text = text.lower()   # lowercase
print("Text length:", len(text), 'characters')
```

```
Text length: 145333 characters
```

```
print(text[:500]) # print first 500 characters
```

```
    alice' s adventures in wonderland | project gutenberg
*** start of the project gutenberg ebook 11 ***   alice' s
adventures in wonderland by lewis carroll the millennium
fulcrum edition 3.0   contents   chapter i.down the rabbit-
hole   chapter ii.the pool of tears   chapter iii.a
caucus-race and a long tale   chapter iv.the rabbit sends
in a little bill   chapter v.advice from a caterpillar
chapter vi.pig and pepper   chapter vii.a mad tea-party
chapter viii.the queen' s croque
```

The simple web scraping code above is at the foundation of code used to build huge text data sets for training of large language models (LLMs).

We can now perform the character count:

```
letter_counts = {}
```

```
for char in text:
# char variable sequentially assumes the value
# of each character in the string variable text
```

```
    # manual check for lowercase alphabetic characters
    # this would return False for punctuation, numbers,
    ↳  uppercase chars, etc.
    if 'a' <= char <= 'z':
```

```
        # key does not exist yet - set its value to zero
        if char not in letter_counts.keys():
            letter_counts[char] = 0

        letter_counts[char] += 1    # increment value by one
        # same as: letter_counts[char] = letter_counts[char]
        ↪  + 1

        # three code lines above could be replaced with one
        ↪  below
        # which automatically returns default 0 value if key
        ↪  does not exist
        # letter_counts[char] = letter_counts.get(char, 0) +
        ↪  1
```

```
letter_counts
```

```
{'a': 8830,
 'l': 4735,
 'i': 7545,
 'c': 2425,
 'e': 13639,
 's': 6515,
 'd': 4944,
 'v': 854,
 'n': 7036,
 't': 10739,
 'u': 3480,
 'r': 5477,
 'w': 2679,
 'o': 8170,
 'p': 1546,
 'j': 149,
 'g': 2540,
 'b': 1488,
 'f': 2005,
 'h': 7399,
```

```
'k': 1161,
'y': 2265,
'm': 2110,
'q': 212,
'x': 152,
'z': 78}
```

Now we just need to sort the result, which can be done using `sorted()` function and `.items()` dictionary function; the latter returns an unsorted list of all (key, value) tuples in a dictionary.

```
sorted_letter_counts = sorted(letter_counts.items(),
                              key=lambda x: x[1],
                              reverse=True)

print("10 most frequent letters:")
for char, freq in sorted_letter_counts[:10]:
    print(f"Letter: '{char}' - Frequency: {freq}")
```

```
10 most frequent letters:
Letter: 'e' - Frequency: 13639
Letter: 't' - Frequency: 10739
Letter: 'a' - Frequency: 8830
Letter: 'o' - Frequency: 8170
Letter: 'i' - Frequency: 7545
Letter: 'h' - Frequency: 7399
Letter: 'n' - Frequency: 7036
Letter: 's' - Frequency: 6515
Letter: 'r' - Frequency: 5477
Letter: 'd' - Frequency: 4944
```

1.15.2 Problem 2 — Most frequent letter bigrams

Now compute 10 most frequent letter bigrams (two-letter combos) in the text we have scraped. (A general bigram or digram is a sequence of two adjacent elements from a string of tokens, which are typically letters, syllables, or words.) Again, treat lower and upper case letters as identical. Only consider adjacent alphabetic characters (ignore numbers, punctuation, etc.).

Solution:

```
bigram_counts = {}
for i in range(len(text) - 1):
    c1, c2 = text[i], text[i + 1]
    if 'a' <= c1 <= 'z' and 'a' <= c2 <= 'z':
        pair = c1 + c2
        bigram_counts[pair] = bigram_counts.get(pair, 0) + 1

top_10_bigrams = sorted(bigram_counts.items(), key=lambda x:
  ↳  x[1], reverse=True)[:10]
print("10 most frequent letter bigrams:")
for pair, freq in top_10_bigrams:
    print(f"Bigram: '{pair}' - Frequency: {freq}")
```

```
10 most frequent letter bigrams:
Bigram: 'he' - Frequency: 3789
Bigram: 'th' - Frequency: 3488
Bigram: 'in' - Frequency: 2028
Bigram: 'er' - Frequency: 1841
Bigram: 'an' - Frequency: 1609
Bigram: 'ou' - Frequency: 1556
Bigram: 'it' - Frequency: 1329
Bigram: 'nd' - Frequency: 1278
Bigram: 'at' - Frequency: 1165
Bigram: 'ha' - Frequency: 1159
```

We could repeat this analysis for longer character sequences, such as trigrams, 4-grams, etc.

1.15.3 Problem 3 — Text cleaning

Write a function to keep only characters and spaces from the raw text of the book – and get rid of punctuation, numbers, etc., replacing them with spaces. Lowercase the letters. Remove any trailing or consecutively repeated spaces.

Solution (a) – manual way:

```python
def get_clean_text(t):
    out = []
    for char in t.lower():
        if ('a' <= char <= 'z'):  # keep
            out.append(char)
        else:
            out.append(" ")

    # turn the list into a string
    out = ''.join(out)

    # split string on spaces and re-join - this gets rid of
    ↪  consecutive spaces
    out = ' '.join(out.split())
    return out.strip()

text_clean = get_clean_text(text)


print(text_clean[:200])
```

```
alice s adventures in wonderland project gutenberg start of
the project gutenberg ebook alice s adventures in wonderland
by lewis carroll the millennium fulcrum edition contents
chapter i down the rab
```

We can also solve this problem via *regular expressions* [Goo24] that offer powerful syntax to match patterns in text – Python provides support for this via package `re`.

Solution (b) – via regular expressions:

```
import re

def get_clean_text(t):
    # replace not lower letters or spaces with spaces
    out = re.sub(r'[^a-z ]', ' ', t.lower())

    # replace any repeated spaces with a single space
    out = re.sub(' +', ' ', out)

    return out.strip()

text_clean = get_clean_text(text)


print(text_clean[:200])
```

```
alice s adventures in wonderland project gutenberg start of
the project gutenberg ebook alice s adventures in wonderland
by lewis carroll the millennium fulcrum edition contents
chapter i down the rab
```

1.15.4 Problem 4 — Most frequent words

Count word frequencies in the cleaned text from the previous problem, omitting any words shorter than two characters. Identify the top 10 most frequent words.

Solution: Here we can just split clean text on spaces (which function as word boundaries) – and drop the 1-character-long elements.

```
words = text_clean.split()                  # split on spaces
words = [w for w in words if len(w) >= 2]   # keep words of 2
  ↪  characters or longer

word_counts = {}
for w in words:
    word_counts[w] = word_counts.get(w, 0) + 1
```

```
sorted_word_counts = sorted(word_counts.items(), key=lambda
    ↪  x: x[1], reverse=True)
top_10_words = sorted_word_counts[:10]

print("10 most frequent words:")
for word, freq in top_10_words:
    print(f"Word: '{word}' - Frequency: {freq}")
```

```
10 most frequent words:
Word: 'the' - Frequency: 1653
Word: 'and' - Frequency: 874
Word: 'to' - Frequency: 729
Word: 'it' - Frequency: 595
Word: 'she' - Frequency: 553
Word: 'of' - Frequency: 517
Word: 'said' - Frequency: 462
Word: 'you' - Frequency: 411
Word: 'alice' - Frequency: 400
Word: 'in' - Frequency: 371
```

1.15.5 Problem 5 — Visualization with Matplotlib

Visualize the 10 most frequent words you just found and their counts using a bar chart. Use `matplotlib` library for plotting.

Solution:

```
# visualization library
import matplotlib.pyplot as plt


# separate words and frequencies
top_words = [w for w, _ in top_10_words]
top_freqs = [f for _, f in top_10_words]

# plotting
```

```
plt.bar(top_words, top_freqs) # bar plot
plt.xlabel("Words")
plt.ylabel("Frequency")
plt.xticks(rotation=45)
plt.savefig('freq_words.pdf')  # save file to pdf
plt.show()
plt.close()
```

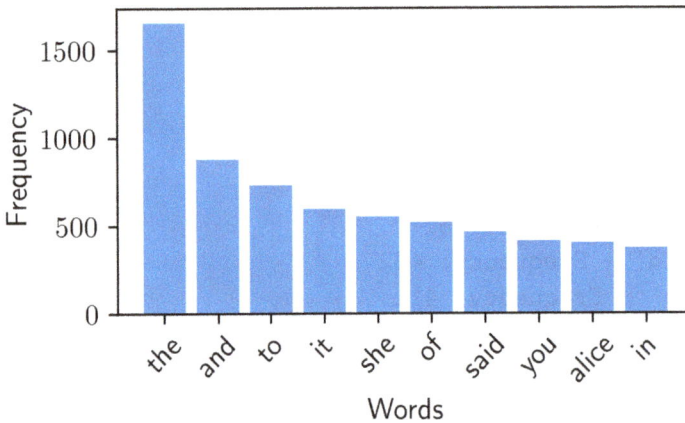

Figure 1.11: Ten most frequent words in "Alice's Adventures in Wonderland"

1.16 Discussion

You have now seen the basics of Python syntax, data types, control flow, and code structuring. (A few more advanced topics – like classes, objects, and decorators – will appear in later chapters.) You have also just used Python to scrape live web data, clean it, analyze it, and visualize the results. This is real data work! You now should hopefully have the foundation to follow the code in the rest of the book on our journey towards LLMs.

1.17 Further learning resources

- Python documentation: https://docs.python.org/.
- Python history: *Python: The Documentary | An origin story* (https://www.youtube.com/watch?v=GfH4QL4VqJ0).
- Software engineer tools (shell scripting, version control / Git, computer networking, etc.): Stanford CS45 course slides (https://web.stanford.edu/class/cs45/).
- Programming abstractions (recursion, data structures, algorithms): Stanford CS106B course (https://web.stanford.edu/class/cs106b/, lectures can be found on YouTube) and [RZ14].
- How computers work at the low level: Stanford CS107 course (https://web.stanford.edu/class/cs107/, lectures can be found on YouTube) and [NS08].
- Web programming with Python and JavaScript: Harvard edX course (https://pll.harvard.edu/course/cs50s-web-programming-python-and-javascript).
- Visualization: Top 50 matplotlib visualizations (https://www.machinelearningplus.com/plots/top-50-matplotlib-visualizations-the-master-plots-python/) and [TG83].
- LeetCode Python challenges: https://leetcode.com/problemset/?search=python.

1.18 References

[Con]   Conda. *Managing environments.* URL: https://docs.conda.io/projects/conda/en/latest/user-guide/tasks/manage-environments.html (visited on 05/16/2025).

[Doc]   Python Docs. *Built-in functions.* URL: https://docs.python.org/3/library/functions.html (visited on 05/02/2025).

[Git24]   GitHub Staff. "Octoverse: AI leads Python to top language as the number of global developers surges". In: *GitHub Blog.* 2024. URL: https://github.blog/news-insights/octoverse/octoverse-2024/.

[Goo24]   Google for Education. *Python Regular Expressions*. 2024. URL: https://developers.google.com/edu/python/regular-expressions (visited on 05/02/2025).

[NS08]   Noam Nisan and Shimon Schocken. *The Elements of Computing Systems: Building a Modern Computer From First Principles*. The MIT Press, 2008.

[Ove]   Stack Overflow. *Why do people write "#!/usr/bin/env python" on the first line of a Python script?* URL: https://stackoverflow.com/questions/2429511/why-do-people-write-usr-bin-env-python-on-the-first-line-of-a-python-script (visited on 05/02/2025).

[Pyt]   Python. *venv – Creation of virtual environments*. URL: https://docs.python.org/3/library/venv.html (visited on 05/16/2025).

[RZ14]   Eric Roberts and Julie Zelenski. *Programming Abstractions in C++*. Pearson, 2014.

[Sol]   Brad Solomon. *Unicode and character encodings in Python: A painless guide*. URL: https://realpython.com/python-encodings-guide/ (visited on 05/16/2025).

[TG83]   Edward R Tufte and Peter R Graves-Morris. *The Visual Display of Quantitative Information*. Graphics Press, 1983.

[Wika]   Wikipedia. *Binary number*. URL: https://en.wikipedia.org/wiki/Binary_number (visited on 05/16/2025).

[Wikb]   Wikipedia. *Bitwise operation*. URL: https://en.wikipedia.org/wiki/Bitwise_operation (visited on 05/16/2025).

[Wikc]   Wikipedia. *History of Python*. URL: https://en.wikipedia.org/wiki/History_of_Python (visited on 08/21/2025).

[Wikd]   Wikipedia. *HTML*. URL: https://en.wikipedia.org/wiki/HTML (visited on 05/02/2025).

[Wike]   Wikipedia. *Real number*. URL: https://en.wikipedia.org/wiki/Real_number (visited on 05/16/2025).

2 Arrays of numbers and their algebra

In this chapter, we review data arrays and operations on them that we need in the rest of the book. We cover (1) linear algebra; (2) NumPy arrays; and (3) Pandas data frames in Python.

2.1 Linear algebra essentials

Most of us start with math on *scalars* – single numbers or functions that output single numbers – such as $2 + 2$, $4 \cdot 2$, or $f(x) = x^2$ (written in Python as `2+2`, `4*2`, `x**2`). Often, though, we need to handle collections of scalars – such as exam grades for n students across k subjects; red, green, blue (RGB) pixel values in an image; and so on. Scalar notation quickly becomes cumbersome for computations on such collections, so we use *linear algebra*, which provides neat ways to work with lists (vectors), tables (matrices), and multidimensional arrays (tensors) of scalars.

> ⚠ Vector, matrix, tensor – terminology differences
>
> In machine learning practice, the terms *vector*, *matrix*, and *tensor* are commonly used to refer to numerical arrays or data containers. For example, a list of 3 numbers – a 1-dimensional or 1D array – is called a vector; a 3×2 (three-by-two) data table – a 2D array – is called a matrix; a $224 \times 224 \times 3$ array of pixel values in an RGB image – a 3D array, as well as a $5 \times 224 \times 224 \times 3$ array representing five RGB images – a 4D array, would both be called tensors. This naming convention is followed by top machine learning libraries in Python, such as PyTorch and TensorFlow. We will also adopt this naming convention and its underlying concrete, numerical view of things.

However, be aware that in abstract mathematics, these same terms – vector, matrix, tensor – refer to more general algebraic / geometric objects, such as elements of vector spaces, linear transformations, or multilinear maps. These objects can be represented by arrays of numbers – but their true meaning goes beyond the specific data representation. For example, an arrow in space (a geometric vector) can be described by different sets of numbers depending on the chosen coordinate system, even though the underlying object – the arrow itself – remains unchanged. There is, admittedly, some tension among professionals well-versed in abstract mathematics related to the redefinition / appropriation of these terms by machine learning people – see [Dis; Staa; Stab; Wike] for discussion.

2.1.1 Matrices and vectors

For illustration, let X be an $n \times k$ matrix of grades ($k = 3$). Each column is one course's grades for n students:

$$X = \begin{bmatrix} x_1 & y_1 & z_1 \\ x_2 & y_2 & z_2 \\ \vdots & \vdots & \vdots \\ x_n & y_n & z_n \end{bmatrix}.$$

Let X_{ij} denote the scalar element in row i and column j of X. For example, element in first row, second column of X is denoted as $X_{12} = y_1$. This denotes some specific grade – say, $y_1 = 96$ points out of 100.

We use the **transpose** operation to flip rows and columns of X, denoted X^T or X'. For example, here:

$$X^T = \begin{bmatrix} x_1 & x_2 & \cdots & x_n \\ y_1 & y_2 & \cdots & y_n \\ z_1 & z_2 & \cdots & z_n \end{bmatrix}.$$

Notice that $X_{ij} = (X^T)_{ji}$.

We will call a single-column $(n \times 1)$ or single-row $(1 \times k)$ matrix a **vector**. For example, a column vector:

$$x = \begin{bmatrix} x_1 \\ x_2 \\ \vdots \\ x_n \end{bmatrix}.$$

And a row vector:

$$x^T = \begin{bmatrix} x_1 & x_2 & \cdots & x_n \end{bmatrix}.$$

We can, for brevity, index vector elements using single i (j) index instead of double ij index.

---

i 2D vs. 1D vectors

To simplify things, we will treat vectors as 2D matrices with a single row or column. If we encounter a 1D array of length n, we will pick whether it is a row or a column, and reshape it into a 2D array of size $1 \times n$ or $n \times 1$ respectively. Once you only work with 2D matrices, all of the usual matrix algebra rules apply cleanly – and there is no need to separately handle special cases for 1D vectors. Keep in mind that some Python libraries do let you keep arrays as 1D, so if you ever feed a 1D array into a matrix operation, you may have to reshape it first for everything to work correctly.

2.1.2 Basic matrix arithmetic

If A and B are the same shape $(n \times k)$:

- **Addition**: $A + B$ is elementwise, so $(A + B)_{ij} = A_{ij} + B_{ij}$.
- **Scalar multiplication**: $c \cdot A$ multiplies each element in A by some number c.
- **Subtraction**: $A - B = A + (-1) \cdot B$.
- **Elementwise (Hadamard) product**: $A \odot B$ multiplies corresponding entries elementwise: $(A \odot B)_{ij} = A_{ij} \cdot B_{ij}$.

Notes:

- When multiplying *scalars* a and b, you will often see $(a \cdot b, a \times b, a * b, ab)$ notation used interchangeably. Python only accepts $*$ though. As notation overload, $\times$ symbol may also serve as a separator for dimension sizes of arrays (e.g., A is $n \times k$ matrix) – this use should be clear from the context.
- Other scalar-matrix operations, e.g., scalar addition, can be defined similarly to scalar multiplication above. For example, $c + A$ adds number c to each element in A.

2.1.3 Broadcasting

Sometimes it is convenient to extend basic arithmetic operations defined above to cases where matrices / arrays do not exactly match in shape but are otherwise somehow compatible. Here are a few common conventions:

- **Single-element array**: Let $X = [x_{11}]$ be a 1×1 matrix (1 row and 1 column) and let A be $n \times k$. We can treat X as functionally equivalent to a scalar when operating on A. For example, elementwise product $[x_{11}] \odot A = x_{11} \cdot A$ or addition $[x_{11}] + A = x_{11} + A$.

  - Technically, $[x_{11}]$ is a container containing a scalar and is not the same as the scalar itself. See more on this nuance here: [Exc].

- **Column and row vectors**: Let c be an $n \times 1$ column vector, r be a $1 \times k$ row vector, and X be an $n \times k$ matrix. Then we can extend elementwise product such that $c \odot X = X \odot c$ means column vector c multiplies each column in X separately elementwise – so $(c \odot X)_{ij} = c_i X_{ij}$; and $r \odot X = X \odot r$ means row vector r multiplies each row in X separately elementwise – so $(r \odot X)_{ij} = r_j X_{ij}$. It is critical here that dimensions align between the vectors and the matrix – length of column vector is n and length of row vector is k for the target $n \times k$ matrix.

More complex broadcasting rules may be supported by specific software libraries – including for higher-dimensional arrays.

2.1.4 Matrix dot product

The **dot product** ($\cdot$) is a very important matrix operation but is a bit more complex in its definition. (It may seem to be a bit artificial when you see it for the first time – and it is because the procedure was concocted by humans for convenience – we will see that it is enormously useful as we progress through the book.)

For a $1 \times k$ row vector l and a $k \times 1$ column vector r, their dot product $l \cdot r$ corresponds to elementwise multiplication of l and r elements with subsequent summation of resulting products – which yields a scalar 1×1 matrix:

$$l \cdot r = \left[\sum_{i=1}^{k} l_i r_i \right] = [l_1 r_1 + l_2 r_2 + \cdots + l_k r_k].$$

> 💡 Σ sum notation
>
> $\sum_{i=1}^{k} l_i r_i$ "Sigma" notation reads "for i taking on integer values from 1 to k (inclusive) sum up all `l_i * r_i`". The expression sets up a summation for loop and is equivalent to the following Python snippet, where l and r are lists of length k (with appropriate change to zero-based indexing):
>
> ```
> sum([l[i]*r[i] for i in range(k)])
> ```

Building on vector dot product, for an $n \times k$ matrix L and a $k \times 1$ vector r, $L \cdot r$ means each row of L is dotted with r, resulting an $n \times 1$ vector (n dot products – one for each row in L).

> 💡 Useful trick – dimension analysis
>
> The dimensionality of the dot product result matrix can be inferred by putting side-by-side the dimensions of the dotted arrays and crossing out the last dimension of the left array and the first dimension of the right array (which must match in size). Here: $L \cdot r \rightarrow (n \times k) \cdot (k \times 1) \rightarrow (n \times \cancel{k}) \cdot (\cancel{k} \times 1) \rightarrow (n \times 1)$ resulting array size. More generally, if L is $n \times k$ and R is $k \times m$, $L \cdot R$ will be $n \times m$, following the dimensionality

computation above $[(n \times k) \cdot (k \times m) \to (n \times \cancel{k}) \cdot (\cancel{k} \times m)]$. Each element of the result is a dot product of one row of L and one column of R. The dot product is only defined when the number of left matrix's columns equals the number of right matrix's rows, so array dimensions must align for the dot product to work.

Example (make sure you understand it!):

$$A = \begin{bmatrix} 1 & 2 & 3 \\ 4 & 5 & 6 \end{bmatrix}, \quad B = \begin{bmatrix} 7 & 8 \\ 9 & 10 \\ 11 & 12 \end{bmatrix}.$$

A is 2×3 matrix, and B is a 3×2 matrix. The resulting matrix $A \cdot B$ will have dimensions $(2 \times 3) \cdot (3 \times 2) \to (2 \times \cancel{3}) \cdot (\cancel{3} \times 2) \to (2 \times 2)$.

Computing the entries of $A \cdot B$, we dot-product each row i of A with each column j of B, recording the result in ijth entry of the resulting matrix. For example, $(A \cdot B)_{11}$ is a dot product of the first row of A and first column of B:

$$A \cdot B = \begin{bmatrix} (1 \cdot 7 + 2 \cdot 9 + 3 \cdot 11) & (1 \cdot 8 + 2 \cdot 10 + 3 \cdot 12) \\ (4 \cdot 7 + 5 \cdot 9 + 6 \cdot 11) & (4 \cdot 8 + 5 \cdot 10 + 6 \cdot 12) \end{bmatrix} = \begin{bmatrix} 58 & 64 \\ 139 & 154 \end{bmatrix}.$$

Transpose of a dot product distributes as follows:

- $(A \cdot B)^T = B^T \cdot A^T$;
- $(A \cdot B \cdot C)^T = C^T \cdot B^T \cdot A^T$;
- and so on.

> **!** Order of dot product
>
> Order of dot product is critical. In general, $A \cdot B$ is not the same as $B \cdot A$. For example, for $1 \times k$ vector l and $k \times 1$ vector r, $l \cdot r$ yields $(1 \times \cancel{k}) \cdot (\cancel{k} \times 1) \to 1 \times 1$ matrix – but, swapping the order, $r \cdot l$ yields $(k \times \cancel{1}) \cdot (\cancel{1} \times k) \to k \times k$ matrix.

As we will see later, neural networks (including large language models) are essentially big chains of dot products. Computer chips called graphical processing units (GPUs) can compute such dot products orders of magnitude faster than regular central processing units (CPUs) and so have played a critical role in the training of large neural nets.

We will now illustrate in Python how such matrix-based notation simplifies code and speeds things up.

For a deeper dive on linear algebra, I recommend these references: [JW07; Cha25].

2.2 NumPy arrays

NumPy is Python's core library for n-dimensional arrays, supporting anything from 1D vectors to 2D matrices or higher-dimensional tensors.

> **i** Keep track of the array shape
>
> NumPy requires us to be explicit about which array type – 1D, 2D, etc. – we are working with. In NumPy, 1D vector is distinct from a 2D matrix with a single row or a single column.

```
import numpy as np
```

2.2.1 Creating arrays

```
np.zeros(6)    # 1D vector of 6 zeros
```

```
array([0., 0., 0., 0., 0., 0.])
```

```
np.ones(5)    # 1D vector of 5 ones
```

```
array([1., 1., 1., 1., 1.])
```

```
np.zeros(6).dtype  # dtype='float64' single float occupies 64
↪  bits in memory
```

```
dtype('float64')
```

```
np.zeros(6).astype('int64') # conversion
```

```
array([0, 0, 0, 0, 0, 0])
```

```
np.zeros((3,4)) # 3x4 matrix of zeros (3 rows, 4 columns)
```

```
array([[0., 0., 0., 0.],
       [0., 0., 0., 0.],
       [0., 0., 0., 0.]])
```

Diagonal matrices:

```
np.eye(3) # 3 by 3 identity - 1 on main diagonal, 0 elsewhere
```

```
array([[1., 0., 0.],
       [0., 1., 0.],
       [0., 0., 1.]])
```

A lower triangular matrix:

```
np.tri(3) # all entries above the main diagonal are zero
```

```
array([[1., 0., 0.],
       [1., 1., 0.],
       [1., 1., 1.]])
```

> **i** Memory vs. numerical precision
>
> In addition to the typical default of 64 bits, NumPy allows floats and integers to be represented with fewer bits, such as 32 or 16. This reduces memory usage and can improve computation speed, but it also limits the range of representable numbers and increases the approximation error between the exact value and its finite-memory representation.
>
> ```python
> np.array([0.123456789], dtype=np.float32)
> ```
>
> ```
> array([0.12345679], dtype=float32)
> ```
>
> ```python
> np.array([0.123456789], dtype=np.float16) # fewer decimal
> ↪ digits are preserved when using 16 bits vs. 32 bits
> ```
>
> ```
> array([0.1235], dtype=float16)
> ```

2.2.2 Example matrix

We create an example array where each row represents a person and each column is a feature (e.g., age, gender, weight):

```python
X = np.array([
    [23, 1, 85],
    [67, 0, 68],
    [43, 0, 100],
    [22, 1, 50]
])
X
```

```
array([[ 23,   1,  85],
       [ 67,   0,  68],
       [ 43,   0, 100],
       [ 22,   1,  50]])
```

We check shape:

```
X.shape
```

```
(4, 3)
```

Get the number of rows – the first number in the shape tuple.

```
X.shape[0]
```

```
4
```

Transpose (rows become columns and vice versa):

```
X.T
```

```
array([[ 23,  67,  43,  22],
       [  1,   0,   0,   1],
       [ 85,  68, 100,  50]])
```

2.2.3 Arithmetic operations

Here are some arithmetic operations on matrices that we discussed earlier:

```
X * 2        # elementwise multiply
```

```
array([[ 46,   2, 170],
       [134,   0, 136],
       [ 86,   0, 200],
       [ 44,   2, 100]])
```

```
X - 1        # elementwise subtract
```

```
array([[22,  0, 84],
       [66, -1, 67],
       [42, -1, 99],
       [21,  0, 49]])
```

```
X ** 2    # elementwise power
```

```
array([[  529,    1,  7225],
       [ 4489,    0,  4624],
       [ 1849,    0, 10000],
       [  484,    1,  2500]])
```

```
X + X    # elementwise addition
```

```
array([[ 46,  2, 170],
       [134,  0, 136],
       [ 86,  0, 200],
       [ 44,  2, 100]])
```

```
X + X.T  # error if shapes don't match
```

```
ValueError: operands could not be broadcast together with
shapes (4,3) (3,4)
```

2.2.4 Indexing

```
X
```

```
array([[ 23,  1,  85],
       [ 67,  0,  68],
       [ 43,  0, 100],
       [ 22,  1,  50]])
```

```
X[0]        # equivalent to X[0, :] - first row
```

```
array([23,  1, 85])
```

```
X[:, 2]    # all rows, 3rd column
```

```
array([ 85,  68, 100,  50])
```

```
X[1,2]     # single element in the second row, third column
```

```
np.int64(68)
```

Boolean indexing:

```
X[:, 1] == 1       # which rows have gender == 1?
```

```
array([ True, False, False,  True])
```

```
X[X[:, 1] == 1]
```

```
array([[23,  1, 85],
       [22,  1, 50]])
```

Assign new values:

```
X
```

```
array([[ 23,   1,  85],
       [ 67,   0,  68],
       [ 43,   0, 100],
       [ 22,   1,  50]])
```

```
X[1,2] = 70 # replace / assign to a specific entry
X
```

```
array([[ 23,    1,   85],
       [ 67,    0,   70],
       [ 43,    0,  100],
       [ 22,    1,   50]])
```

```
X[:,1] = 0   # replace the second column with zeros
X
```

```
array([[ 23,    0,   85],
       [ 67,    0,   70],
       [ 43,    0,  100],
       [ 22,    0,   50]])
```

```
X[:, 1] = [1, 0, 0, 1]   # replace the second column with a
 ↪   list of numbers
X
```

```
array([[ 23,    1,   85],
       [ 67,    0,   70],
       [ 43,    0,  100],
       [ 22,    1,   50]])
```

2.2.5 Speed of NumPy vs. pure Python loops

Consider adding 2 to each element of a 10-million-element array:

```
my_list = list(range(10000000))    # simple list
my_arr = np.arange(10000000)        # numpy array
```

```
%time for i in range(10): [2 + x for x in my_list]
```

```
CPU times: user 1.93 s, sys: 340 ms, total: 2.27 s
Wall time: 2.27 s
```

```
%time for i in range(10): 2 + my_arr
```

```
CPU times: user 36.5 ms, sys: 46.6 ms, total: 83.1 ms
Wall time: 83.2 ms
```

Vector-based NumPy implementation is orders of magnitude faster than the for loop!

---

i Timing the computation

`%time` is a magic function for timing computation in Jupyter – but it would not work in raw Python. Alternatively, we could use the following timing structure, which would work across the board:

```python
import time
start = time.time()
# computation...
end = time.time()
print(end - start, "seconds")
```

Some of the reasons for this include more efficient memory layout for NumPy arrays, NumPy's highly efficient underlying C language implementation, and even NumPy potentially relying on hardware-optimized linear algebra arithmetic on user's machine via BLAS (Basic Linear Algebra Subprograms [Wika]).

Vectorizing your code (replacing for loops with matrix operations) can thus reduce runtime substantially when it comes to Python. This becomes particularly critical for long-running computations – where vectorization can be a difference between hours and days of compute! You can read more about this topic here [Ovea; Ovec].

> **i** Fast for loops
>
> Not all languages suffer a performance penalty for explicit loops vs. vectorized code: for example, Julia's compiler makes plain for loops as fast as its vectorized operations, which can be a huge convenience when the vectorization is impractical (e.g., a problem is hard to vectorize). Compilers – programs that translate high-level code into optimized machine instructions – are key to these speed gains. In Python, you can close the speed gap in two main ways. First, just-in-time (JIT) compilers like `numba` can produce optimized machine code from Python code, often delivering C-like speeds with minimal code changes. Second, depending on the problem's structure, it may be possible to parallelize its execution, splitting work across multiple workers (threads / processes / machines) in a way that delivers significant speedups. Many libraries exist for this – `multiprocessing` or `threading` for local CPU parallelism, `PySpark` or `Dask` for clusters / distributed computing, or GPU-accelerated frameworks like `PyTorch` .

2.2.6 Dot product

```
X
```

```
array([[ 23,   1,  85],
       [ 67,   0,  70],
       [ 43,   0, 100],
       [ 22,   1,  50]])
```

Remember – dot product requires matching dimensions:

```
X.dot(X) # fails - num of columns in left matrix != num of
↪ rows in second matrix
```

```
ValueError: shapes (4,3) and (4,3) not aligned: 3 (dim 1) !=
4 (dim 0))
```

```
XtX = X.T.dot(X) # this works: (3 by 4) dot (4 by 3); .T is
↳  transpose
XtX.shape
```

```
(3, 3)
```

```
XtX
```

```
array([[ 7351,     45, 12045],
       [   45,      2,   135],
       [12045,    135, 24625]])
```

Alternative ways to run dot product, equivalent to `X.T.dot(X)` :

```
X.T @ X
```

```
array([[ 7351,     45, 12045],
       [   45,      2,   135],
       [12045,    135, 24625]])
```

```
np.dot(X.T, X)
```

```
array([[ 7351,     45, 12045],
       [   45,      2,   135],
       [12045,    135, 24625]])
```

2.2.7 Matrix inverse

In case of square matrices (number of rows and columns is the same, e.g., $A = X^TX$ from above), we can ask, what square matrix B could we multiply A by so that the result of the dot product is the identity matrix $I = A \cdot B = B \cdot A$, which is also square and has ones in its primary diagonal (ii indices for $i = 1, 2, ...$) and zeros everywhere else. Such a matrix B is called an inverse of A and is denoted as A^{-1}. This generalizes the idea of inverse scalars. There exist rather tedious methods for finding such inverses (see [Wikc]) – we will use NumPy matrix inverse implementation as follows:

```
inv = np.linalg.inv(XtX)
inv
```

```
array([[ 9.02365638e-04,  1.50646344e-02, -5.23968315e-04],
       [ 1.50646344e-02,  1.04521281e+00, -1.30987716e-02],
       [-5.23968315e-04, -1.30987716e-02,  3.68711980e-04]])
```

```
np.dot(XtX, inv).round(5)  # we can verify inv is accurate up
↪  to some precision
```

```
array([[ 1.,  0., -0.],
       [ 0.,  1.,  0.],
       [ 0.,  0.,  1.]])
```

ℹ Advanced optional material: Invertibility

If the columns or rows of a matrix are linearly dependent, the matrix has no inverse. This means that at least one column (or row) can be written exactly as a linear combination of others (e.g., $z = 0.1x - 0.5y$, where x, y, and z are columns). The computation of a matrix inverse involves division by a quantity called the *determinant*. If the determinant is zero, the inverse is undefined – this happens precisely when the matrix has linearly dependent columns or rows. Geometrically, the determinant represents the volume of the shape (e.g., a parallelogram in 2D, a parallelepiped in 3D) spanned / outlined by the matrix's columns or rows. When these vectors are linearly dependent, the shape collapses into a lower dimension, and the volume becomes zero (like a parallelepiped in 3D with all its facets flattened, parallel, contained within a 2D slice of the 3D space). In practical applications, this can lead to numerical issues. For example, if a column is duplicated in a data set, it introduces exact linear dependence, making the matrix non-invertible. Even if columns are not exactly the same but are highly correlated, the determinant becomes very small, making the inverse numerically unstable. This condition is known as *multicollinearity* in statistics and machine learning.

2.2.8 Some useful matrix functions

```
X
```

```
array([[ 23,    1,   85],
       [ 67,    0,   70],
       [ 43,    0,  100],
       [ 22,    1,   50]])
```

```
np.sum(X, 0)   # sum of each column
```

```
array([155,    2,  305])
```

```
np.sum(X, 1)   # sum of each row
```

```
array([109, 137, 143,   73])
```

```
np.mean(X, 0)  # avg. of each column
```

```
array([38.75,   0.5 ,  76.25])
```

```
np.sum(X, 0) / X.shape[0]  # avg. of each column
```

```
array([38.75,   0.5 ,  76.25])
```

```
np.std(X, 0)
# standard deviation of each column
# more on this in a later section
```

```
array([18.33541655,  0.5        ,  18.49831073])
```

Functions above can be implemented via dot product operations.

For example, we can use dot product to get sum of elements in a matrix. To sum elements in each column, we could dot product each column with a vector containing ones (each column element gets multiplied by one and all such products are added up by dot product definition). For example, 1×4 2D matrix of ones dot 4×3 2D matrix yields 1×3 vector:

```
out = np.ones((1,4)).dot(X)
print(out)
print(out.shape)
```

```
[[155.   2. 305.]]
(1, 3)
```

Additionally, NumPy supports dot product between 1D vector and 2D matrix – 1D vector of length 4 dot 2D 4×3 matrix yields a 1D vector of length 3 – so the result is similar to one above but the first singleton dimension is not involved:

```
out = np.ones(4).dot(X)    # same as np.sum(X, 0)
print(out)
print(out.shape)
```

```
[155.   2. 305.]
(3,)
```

Functions based on the ordering of data, such as the maximum or median, are also commonly used in data analysis. For example, maximum returns the largest value:

```
np.max(X, 0) # maximum of each column
```

```
array([ 67,   1, 100])
```

Median returns the value that splits the data such that 50% of the values are greater and 50% of the data are smaller than that median value. For a sorted list of values, if the number of elements is odd, the middle element is the median. If the number of elements is even, the median is usually computed as the average of the two middle elements.

```
np.median(X, 0) # median of each column
```

```
array([33. ,   0.5, 77.5])
```

The arithmetic mean we saw earlier (sum of values divided by their number) and the median are two measures of the "center" of the data. Median is considered a more *robust* measure, because it is not affected by an occasional outlier. Consider 1-5 data series containing a typo due to input error (number 40 instead of number 4):

```
np.mean([1,2,3,40,5]) # mean is affected by the outlier
```

```
np.float64(10.2)
```

```
np.median([1,2,3,40,5]) # median ignores the outlier
```

```
np.float64(3.0)
```

Percentile function generalizes different measures based on sorted data – it returns the value below which a given percentage of observations in the data falls:

```
p = np.percentile([1,2,3,4,5], q=75, method='linear')
p
# ~75% of data can be expected to fall below this value
# there exist different methods to compute the percentile -
↪ google the docs
```

```
np.float64(4.0)
```

> 💡 NumPy scalars to Python scalars
>
> If desired, we can convert NumPy scalars to Python scalars using
> `.item()` function:
>
> ```
> p.item() # NumPy scalar to Python scalar
> ```
>
> ```
> 4.0
> ```

2.2.9 Handling singleton dimensions

As discussed earlier, we can represent a list of numbers as a 1D vector or as 2D column or row vectors. In case of 2D vectors, singleton dimensions arise – for example, the first dimension of a row vector of size 1×3 is called singleton – the vector contains three numbers and the singleton dimension explicitly indicates the numbers are arranged as a row of a matrix.

Sometimes, it may be necessary to move between 2D and 1D structures – getting rid of or adding singleton dimensions. Here is how we can do it:

```
v = np.ones((1,3))
print(v)
print(v.shape)
```

```
[[1. 1. 1.]]
(1, 3)
```

We can squeeze out length-1 dimensions, converting 2D row vector into a 1D vector:

```
v = np.squeeze(v)
print(v)
print(v.shape)
```

```
[1. 1. 1.]
(3,)
```

Or expand dims:

```
v = np.ones(3)
print(v)
print(v.shape)
```

```
[1. 1. 1.]
(3,)
```

```
out = np.expand_dims(v,0)
print(out) # row vector
print(out.shape)
```

```
[[1. 1. 1.]]
(1, 3)
```

```
out = np.expand_dims(v,1)
print(out) # column vector
print(out.shape)
```

```
[[1.]
 [1.]
 [1.]]
(3, 1)
```

2.2.10 Reshaping

Sometimes we may need to change the shape of the matrix while preserving the number of elements. However, if we change the shape, how are elements rearranged? To answer this, let us first flatten a matrix to a 1D vector.

```
X
```

```
array([[ 23,   1,  85],
       [ 67,   0,  70],
       [ 43,   0, 100],
       [ 22,   1,  50]])
```

```
print(X.flatten())  # row-major flatten
```

```
[ 23   1  85  67   0  70  43   0 100  22   1  50]
```

We can observe that the matrix is converted to 1D vector by sequentially concatenating rows. This is called row-major order [Wikd]. In fact, the flattened version of the matrix is how data is actually stored in memory – elements in rows contiguously following each other. This is in contrast to column major order, where elements in columns are contiguously collocated in memory.

This is useful to know for two reasons. First, in some scenarios, there can be significant speed implications of retrieving or writing data in large matrices along rows vs. columns – if row index is more frequently accessed, row major order would be more advantageous. Second, more importantly here, knowing that matrix stores data in row major order allows us to predict how content is arranged in a reshaped matrix.

```
X.reshape(3,4)  # populated row-wise from X.flatten()
```

```
array([[ 23,   1,  85,  67],
       [  0,  70,  43,   0],
       [100,  22,   1,  50]])
```

Note that **reshape** function can also be used to handle singleton dimensions.

2.2.11 Broadcasting

```
X
```

```
array([[ 23,    1,   85],
       [ 67,    0,   70],
       [ 43,    0,  100],
       [ 22,    1,   50]])
```

NumPy "broadcasts" arrays of compatible shapes. For instance, we can subtract column means from each row:

```
np.mean(X, 0) # within-column means - 1D vector
```

```
array([38.75,   0.5 ,  76.25])
```

```
X - np.mean(X, 0)
```

```
array([[-15.75,    0.5 ,    8.75],
       [ 28.25,   -0.5 ,   -6.25],
       [  4.25,   -0.5 ,   23.75],
       [-16.75,    0.5 ,  -26.25]])
```

To subtract row means from each column, however, we explicitly need them in a 4×1 shape:

```
X - np.mean(X, 1) # error - because wrong shape of the vector
```

```
ValueError: operands could not be broadcast together with
shapes (4,3) (4,)
```

```
X - np.mean(X, 1).reshape(4,1) # this works
```

```
array([[-13.33333333, -35.33333333,  48.66666667],
       [ 21.33333333, -45.66666667,  24.33333333],
       [ -4.66666667, -47.66666667,  52.33333333],
       [ -2.33333333, -23.33333333,  25.66666667]])
```

Adding 1×1 matrix is also a form of broadcasting NumPy supports:

```
X + np.ones((1,1))
```

```
array([[ 24.,   2.,   86.],
       [ 68.,   1.,   71.],
       [ 44.,   1.,  101.],
       [ 23.,   2.,   51.]])
```

2.3 Application: Covariance and correlation matrices

Covariance quantifies how two variables change together:

- A *positive* covariance means that as one variable increases, the other tends to increase as well (e.g., on average, education and income [BEN15]).
- A *negative* covariance means that when one variable increases, the other tends to decrease (e.g., on average, the number of cigarettes smoked per day and longevity [Str+07]).
- The covariance of a variable with itself is called its **variance**, which measures how spread out its values are around the mean. A larger variance indicates greater dispersion.

The *sample covariance* between two vectors a and b, each of length n, is computed as:

$$\text{Cov}(a, b) = \frac{1}{n-1} \sum_{i=1}^{n} (a_i - \bar{a})(b_i - \bar{b}),$$

where $\bar{a} = \frac{1}{n}\sum_{i=1}^{n} a_i$ and $\bar{b} = \frac{1}{n}\sum_{i=1}^{n} b_i$ are the sample means (averages) of the vectors.

Here is an example calculation of covariance between two different variables:

```python
# data
a = np.array([2, 4, 6, 8])
b = np.array([1, 3, 5, 7])
n = len(a)

# compute the means
mean_a = np.sum(a) / n
mean_b = np.sum(b) / n

# sum up covariance across observations in the data
cov_sum = 0
for i in range(n):
    cov_sum += (a[i] - mean_a) * (b[i] - mean_b)
    # if both values for an observation are higher (or lower)
    ↪    than the average
    # then the product is positive - e.g., income and
    ↪    education of a person
    # that is - variables go in the same direction relative
    ↪    to the mean
    # if one deviation from the mean is positive, and another
    ↪    is negative
    # then the product is negative - e.g.,
    # daily number of cigarettes smoked and longevity
    # that is - variables go in different directions vs. mean

# divide by (n - 1) to get the sample covariance
sample_cov = cov_sum / (n - 1)

print("Sample covariance:", sample_cov)


Sample covariance:  6.666666666666667
```

> **i** *n* vs. *n* − 1
>
> The term *sample* in sample covariance reflects the fact that we usually compute it from a subset of data, not the entire population (e.g., survey sample vs. all USA population). We divide by $n-1$ instead of n (known as *Bessel's correction*) because we use our data to estimate both the mean and the deviations from the mean, which underestimates data variability with respect to the true overall population mean that we do not actually observe. When n is large, numerical differences resulting from n vs. $n-1$ use become negligible. For a more detailed explanation, see [Wikb].

Covariance of a vector with itself is its variance – average of squared deviations from the mean:

$$\text{Var}(a) = \text{Cov}(a, a) = \frac{1}{n-1} \sum_{i=1}^{n} (a_i - \bar{a})(a_i - \bar{a}) = \frac{1}{n-1} \sum_{i=1}^{n} (a_i - \bar{a})^2.$$

A **standard deviation** is a square root of the variance:

$$\text{SD}(a) = \sqrt{\text{Var}(a)}.$$

Here is an example of variance calculation in code (same result is we set $b = a$ in the previous code snippet):

```
# data
a = np.array([2, 4, 6, 8])
n = len(a)

# compute the means
mean_a = np.sum(a) / n

# sum up covariance across data observations
var_sum = 0
for i in range(n):
    var_sum += (a[i] - mean_a)**2

# divide by (n - 1) to get the sample variance
sample_var = var_sum / (n - 1)
```

```
# take square root
sample_sd = np.sqrt(sample_var)
print("Sample variance:", sample_var)
print("Sample standard deviation:", sample_sd)
```

```
Sample variance: 6.666666666666667
Sample standard deviation: 2.581988897471611
```

For verification:

```
np.std(a, ddof = 1) # 1 in divisor n-1 is called degrees of
↪   freedom
```

```
np.float64(2.581988897471611)
```

A **covariance matrix** of size $k \times k$ captures the *covariance* between every pair of variables in a data set of k variables. The ijth entry of this matrix records covariance between variables i and j. The main diagonal, where $i = j$, records the variances – covariance of each variable with itself.

We can compute the sample covariance matrix of the columns in X using simple matrix expressions:

```
X
```

```
array([[ 23,    1,   85],
       [ 67,    0,   70],
       [ 43,    0,  100],
       [ 22,    1,   50]])
```

```
Xc = X - np.mean(X, 0) # centering columns (subtract means)
S = Xc.T.dot(Xc) / (Xc.shape[0] - 1)
S
```

```
array([[ 4.48250000e+02, -1.08333333e+01,  7.54166667e+01],
       [-1.08333333e+01,  3.33333333e-01, -5.83333333e+00],
       [ 7.54166667e+01, -5.83333333e+00,  4.56250000e+02]])
```

In a one-line operation, matrix dot product automatically computes a dot product of every column of X with every column of X.

To verify correctness:

```
np.cov(X, rowvar=False)    # columns are variables
```

```
array([[ 4.48250000e+02, -1.08333333e+01,  7.54166667e+01],
       [-1.08333333e+01,  3.33333333e-01, -5.83333333e+00],
       [ 7.54166667e+01, -5.83333333e+00,  4.56250000e+02]])
```

If – in addition to centering (subtracting the mean) – we also divide input variables by their *standard deviations*, the resulting covariance matrix becomes a **correlation matrix** – and its values strictly lie between −1 and 1:

$$\text{Corr}(a, b) = \frac{1}{n-1} \sum_{i=1}^{n} \left[\frac{(a_i - \bar{a})}{\sqrt{\text{Var}(a)}} \frac{(b_i - \bar{b})}{\sqrt{\text{Var}(b)}} \right].$$

This normalization facilitates comparability across variables with different units or scales:

- Correlation values close to 1 signal a very strong positive association – as one variable increases, so does the other.
- Values near −1 signal a strong negative association – as one variable increases, the other tends to decrease.

In code, we can compute the sample correlation matrix from a data matrix X using:

```
Xc = X - np.mean(X, 0) # centering columns (subtract means)
var = np.sum(Xc**2, 0) / (X.shape[0] - 1) # sample variances
sd = np.sqrt(var) # standard deviations
Xn = Xc / sd  # dividing centered columns by their standard
↪   deviations
```

```
corr = Xn.T.dot(Xn) / (Xn.shape[0] - 1) # covariance on
 ↪ normalized columns
corr
```

```
array([[ 1.        , -0.88626293,  0.16676532],
       [-0.88626293,  1.        , -0.47301616],
       [ 0.16676532, -0.47301616,  1.        ]])
```

To verify correctness:

```
np.corrcoef(X, rowvar=False)   # columns are variables
```

```
array([[ 1.        , -0.88626293,  0.16676532],
       [-0.88626293,  1.        , -0.47301616],
       [ 0.16676532, -0.47301616,  1.        ]])
```

Both covariance and correlation metrics presented here measure *linear* as-sociation – they may not capture an association that is *non-linear* (e.g., one variable varies like a sine wave with respect to values of another vari-able). So correlation and covariance values close to 0 suggests little or no *linear* association between the variables – even though a complex non-linear dependence may exist.

The presented correlation quantity is more formally known as the *Pearson correlation coefficient*. Other types of correlation coefficients can be defined, such as rank correlation coefficients (Kendall's Tau, Spearman's, etc.), which are less sensitive to outliers and are better at capturing some types of non-linear association in the data.

2.4 Sidebar: Exponential and logarithmic functions

We will seize here an opportunity to discuss exponential and logarithmic functions, which are fundamental in machine learning.

2.4.1 Exponential function

The **exponential function**, written as $\exp(x)$ or e^x, raises Euler's constant $e \approx 2.71828$ to the power of input x. In NumPy, it is implemented as `np.exp()` and applies elementwise when used on an array.

```
X
```

```
array([[ 23,   1,   85],
       [ 67,   0,   70],
       [ 43,   0,  100],
       [ 22,   1,   50]])
```

```
Y = np.exp(X)   # elementwise exponentiation 2.718..^X
Y
```

```
array([[9.74480345e+09, 2.71828183e+00, 8.22301271e+36],
       [1.25236317e+29, 1.00000000e+00, 2.51543867e+30],
       [4.72783947e+18, 1.00000000e+00, 2.68811714e+43],
       [3.58491285e+09, 2.71828183e+00, 5.18470553e+21]])
```

More examples:

```
np.exp(1)   # 2.718..^1 = 2.718
```

```
np.float64(2.718281828459045)
```

```
np.exp(0)   # 2.718..^0 = 1.0
```

```
np.float64(1.0)
```

```
np.exp(-1)   # 2.718..^{-1} = 1/2.718
```

```
np.float64(0.36787944117144233)
```

Notes on `exp` function:

- Exponentiation maps a *real* number x – a number that ranges from negative infinity to plus infinity $x \in (-\infty, \infty)$ – to strictly positive values, so $e^x \in (0, \infty)$.
- $\exp(x + y) = \exp(x) \cdot \exp(y)$.
- The exponential function equals its own derivative – we will discuss derivatives in later chapters.
- `exp` is widely used in machine learning (e.g., softmax, log-likelihood functions – we will discuss these later).
- Some sources use the name *exponential* for general power functions of the form a^x, where a is a constant. In this book, we will reserve the term *exponential function* specifically for those with Euler's constant e as the base – that is, the *natural* exponential functions.

2.4.2 Natural logarithm

As a reminder, a logarithm function with *base* a and input x returns a power to which we need to bring a to get x: $a^y = x \iff \log_a(x) = y$. For example, $\log_{10}(1000) = 3$ because $10^3 = 1000$.

Natural logarithm is a logarithm function with Euler's number ($e \approx 2.71828$) as the base:

$$\log_e(x) = y \iff e^y = x.$$

You may encounter different equivalent math notation for the natural logarithm: $\ln(x) = \log(x) = \log_e(x)$. In this book, like in many areas of science, when the base of the algorithm is not specified, we will assume we are dealing with the natural logarithm.

The natural logarithm definition means it is the inverse of the exponential function: $\log(\exp(x)) = x$ for $x \in (-\infty, \infty)$, and $\exp(\log(x)) = x$ for $x > 0$.

The natural logarithm requires its input x to be positive – in the $(0, \infty)$ range (because we cannot get a negative number by bringing a positive number to a power). Also note that as input approaches zero $x \to 0$, the logarithm goes to negative infinity: $\lim_{x \to 0} \log_e x \to -\infty$.

In Python, we calculate the natural logarithm using `np.log()` :

```
X
```

```
array([[ 23,    1,    85],
       [ 67,    0,    70],
       [ 43,    0,   100],
       [ 22,    1,    50]])
```

```
Y = np.exp(X)  # elementwise exponentiation 2.718..^X
Y
```

```
array([[9.74480345e+09, 2.71828183e+00, 8.22301271e+36],
       [1.25236317e+29, 1.00000000e+00, 2.51543867e+30],
       [4.72783947e+18, 1.00000000e+00, 2.68811714e+43],
       [3.58491285e+09, 2.71828183e+00, 5.18470553e+21]])
```

```
np.log(Y)  # elementwise natural logarithm
```

```
array([[ 23.,    1.,    85.],
       [ 67.,    0.,    70.],
       [ 43.,    0.,   100.],
       [ 22.,    1.,    50.]])
```

In general, $\log(0)$ is undefined. However, sometimes, by a relatively common mathematical convention, we can agree that $\exp(-\infty) = 0$ and $\log(0) = -\infty$, extending input domain of log to zero values. This convention is also followed in NumPy (although it may emit runtime warnings for dangerous operations):

```
np.exp(-np.inf)
```

```
np.float64(0.0)
```

```
np.log(0)
```

```
divide by zero encountered in log
```

```
-inf
```

np.log of a negative number will return **nan** .

```
np.log(-1)
```

```
invalid value encountered in log
```

```
nan
```

Base 10 logarithm. Other logarithm bases are common too – for example, logarithm with base 10, $\log_{10}(x)$:

```
np.log10(Y)  # elementwise base 10 logarithm
```

```
array([[ 9.98877308,  0.43429448, 36.91503096],
       [29.09773029,  0.        , 30.40061373],
       [18.67466272,  0.        , 43.42944819],
       [ 9.5544786 ,  0.43429448, 21.7147241 ]])
```

This tells you roughly how many zeros a number has – how many times 10 must be multiplied by itself to reach that number. For example:

```
np.log10(1000) # 3
```

```
np.float64(3.0)
```

Such logarithmic base 10 transformation is popular when handling price data (large positive numbers, outlier values are common). In this case, the log-transformed values will have much narrower range than raw values – for example, $\log_{10}(\$10) = 1$ and $\log_{10}(\$1,000,000,000) = 9$ – we will see an example later in the chapter. If the price data contains zeros, one can either drop those observations or add a small positive constant (e.g., 1) before taking the logarithm (to avoid the logarithm of zero); this introduces a slight bias – however, it is negligible when prices are large.

Logarithmic identities. Recall that logarithm (for all positive bases $a \neq 1$) possesses a highly useful property:

- Logarithm converts product to addition: $\log(x \cdot y) = \log(x) + \log(y)$. More generally, $\log(\prod_{i=1}^{n} x_i) = \sum_{i=1}^{n} \log x_i$. Here $\prod_{i=1}^{n} x_i$ means "multiply all x_i for $i = 1, 2, ..., n$ together" and is equivalent to the loop:

```
product = 1.0
for x_i in x:
    product *= x_i # same as product = product * x_i
```

- By extension, logarithm converts the *taking to the power* to a *product with a scalar*: $\log(x^y) = y \cdot \log(x)$. For example, $\log(x^3) = \log(x \cdot x \cdot x) = \log(x) + \log(x) + \log(x) = 3\log(x)$. By implication, $\log(\frac{1}{x}) = \log(x^{-1}) = -\log(x)$. Thus, $\log(x/y) = \log(x \cdot y^{-1}) = \log(x) - \log(y)$.

These identities are *crucial* when dealing with products of small or large values. For example, multiplying multiple probabilities – typically, small numbers – can cause an underflow error (a calculation results in a value smaller than the smallest representable value for a given data type), but adding the logs of small numbers preserves numerical stability.

For instance, consider taking logarithm of a product of small quantities:

```
tiny_values = np.full(1000, 1e-5)  # 1000-element array
 ↪   filled with 0.00001
# 1e-5 is scientific notation for 0.00001 - and is also a
 ↪   float
# here, by convention, e stands for 10, not Euler's constant
 ↪   though...
np.prod(tiny_values)  # product of array elements
```

```
np.float64(0.0)
```

We get a silent underflow error in the product as NumPy rounds the non-zero result to a zero.

Taking log of this yields a negative infinity.

```
np.log(np.prod(tiny_values))
```

```
divide by zero encountered in log
```

```
-inf
```

In contrast, if we use the logarithmic identity and take log of the array elements first and then sum up the values, we get the correct numerical result:

```
np.sum(np.log(tiny_values))
```

```
np.float64(-11512.925464970229)
```

Finally, let us plot **exp** and **log** functions like this:

```python
import matplotlib.pyplot as plt
```

```python
# plotting
x = np.linspace(0.01, 2.5, 500) # equally spaced 500 points
↪ between specified numbers
plt.plot(x, np.exp(x), label='exp(x)', linestyle='--',
↪ linewidth=2)
plt.plot(x, np.log(x), label='ln(x)', linewidth=2)
plt.xlabel("x")
plt.ylabel("f(x)")
plt.legend()
plt.grid(True, linestyle=':', linewidth=0.5)
plt.savefig("log_exp_plot.png")
plt.show()
plt.close()
```

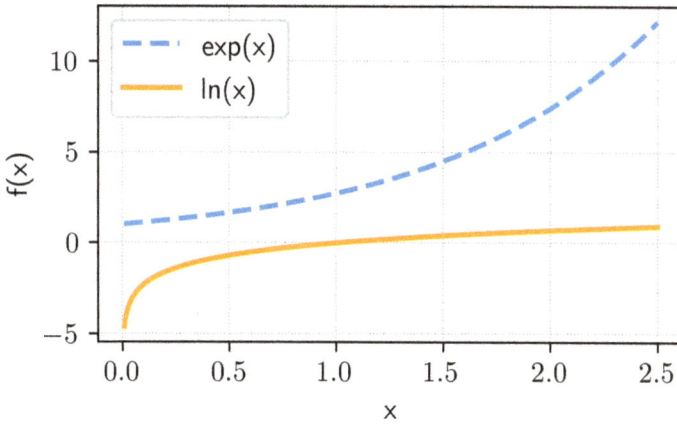

Figure 2.1: Exp and Log functions

2.5 Pandas data tables

Pandas is a powerful library in Python that enables loading, cleaning, and processing data in table format. It interacts nicely with basic Python structures such as dictionaries and lists and with NumPy arrays. We will see here a few examples of its use – first on toy data and then based on some real New York City real estate data.

The core tool in Pandas is the DataFrame object – you can think of it as a table with a lot of useful functions and properties. Here is how to form a Pandas data frame from a dictionary.

```python
import pandas as pd
d = pd.DataFrame({
    "Name": ["Yegor", "Sasha", "Bob", "Alex"],
    "Age": [30,28,42,22],
    "Well Traveled": [True, False, True, False]
})
d
```

	Name	Age	Well Traveled
0	Yegor	30	True
1	Sasha	28	False
2	Bob	42	True
3	Alex	22	False

Get some important properties from the DataFrame object:

```
d.columns
```

```
Index(['Name', 'Age', 'Well Traveled'], dtype='object')
```

```
d.index
```

```
RangeIndex(start=0, stop=4, step=1)
```

```
d["Age"].values
```

```
array([30, 28, 42, 22])
```

Indexing by position vs. label:

```
# positional index
d.iloc[2, :]
```

```
Name           Bob
Age             42
Well Traveled  True
Name: 2, dtype: object
```

```
d.iloc[:, 1]
```

```
0    30
1    28
2    42
3    22
Name: Age, dtype: int64
```

```
# name-based indexing
d.loc[:, 'Age']
```

```
0    30
1    28
2    42
3    22
Name: Age, dtype: int64
```

Multiple columns:

```
d.loc[:, ["Name", 'Age']]
```

	Name	Age
0	Yegor	30
1	Sasha	28
2	Bob	42
3	Alex	22

Concatenate:

```
pd.concat([d,d],axis=0).reset_index(drop=True)
```

	Name	Age	Well Traveled
0	Yegor	30	True
1	Sasha	28	False
2	Bob	42	True
3	Alex	22	False
4	Yegor	30	True
5	Sasha	28	False
6	Bob	42	True
7	Alex	22	False

Merging:

```
d1 = pd.DataFrame({"Name" : ["Yegor", "Sasha"],
                   "Height": [192, 168] })
d1
```

	Name	Height
0	Yegor	192
1	Sasha	168

```
d.merge(d1, how='inner', left_on = "Name", right_on="Name")
```

	Name	Age	Well Traveled	Height
0	Yegor	30	True	192
1	Sasha	28	False	168

```
d.merge(d1, how='left', left_on = "Name", right_on="Name")
```

	Name	Age	Well Traveled	Height
0	Yegor	30	True	192.0

	Name	Age	Well Traveled	Height
1	Sasha	28	False	168.0
2	Bob	42	True	NaN
3	Alex	22	False	NaN

```
np.nan  # this is nan value appearing in pandas for missing
↪ values
```

```
nan
```

2.6 Data exercise

The following exercise is inspired by an assignment from Columbia University's "Introduction to Data Science" course, offered back in 2012 by Dr. Rachel Schutt and co-taught by Dr. Cathy O'Neil and Jared P. Lander – a class the author of this book had the pleasure of taking. Two of the instructors later published a book [OS13] based on this course.

In this exercise, we will download real estate transaction data for the New York City (NYC), read it into a Pandas data frame, clean the data, and do basic analysis and visualizations.

```
url = ("https://www1.nyc.gov/assets/finance/downloads/pdf"
        "/rolling_sales/rollingsales_manhattan.xlsx")
# this is a way to write a single string on multiple lines
# using brackets; note no commas, or it would become a tuple

# download and save the excel file with python
# note that this data will change over time
import requests
response = requests.get(url)
with open('rollingsales_manhattan.xlsx', "wb") as f:
    f.write(response.content)
```

```
# column names start on fifth row, so we skip the first 4
data = pd.read_excel("rollingsales_manhattan.xlsx",
 ↪  skiprows=4)
print(data.head(1).T) # show first 1 row (transposed)
```

	0
BOROUGH	1
NEIGHBORHOOD	ALPHABET CITY
BUILDING CLASS CATEGORY	01 ONE FAMILY DWELLINGS
TAX CLASS AT PRESENT	1
BLOCK	376
LOT	43
EASEMENT	NaN
BUILDING CLASS AT PRESENT	S1
ADDRESS	743 EAST 6 STREET
APARTMENT NUMBER	NaN
ZIP CODE	10009
RESIDENTIAL UNITS	1.0
COMMERCIAL UNITS	1.0
TOTAL UNITS	2.0
LAND SQUARE FEET	2090.0
GROSS SQUARE FEET	3680.0
YEAR BUILT	1940.0
TAX CLASS AT TIME OF SALE	1
BUILDING CLASS AT TIME OF SALE	S1
SALE PRICE	0
SALE DATE	2025-01-23 00:00:00

```
data.shape
```

```
(18253, 21)
```

```
# first three column names
list(data.columns[:3])
```

```
['BOROUGH', 'NEIGHBORHOOD', 'BUILDING CLASS CATEGORY']
```

Clean data:

```
data = data[~data["YEAR BUILT"].isna()] # year is not
 ↪ missing, tilde ~ inverts boolean values
data = data[data['YEAR BUILT'] >= 1900] # year >= 1900
data = data[data['SALE PRICE'] > 100000] # high enough price
data.shape
```

```
(12392, 21)
```

```
data['TAX CLASS AT TIME OF SALE'].value_counts()
```

```
TAX CLASS AT TIME OF SALE
2    11627
4      609
1      156
Name: count, dtype: int64
```

```
data['TAX CLASS COMMERCIAL'] = 1.0*(data['TAX CLASS AT TIME
 ↪ OF SALE'] == 4)
data.to_csv('clean_data.csv', index=False) # save as .csv
# data = pd.read_csv('clean_data.csv')
```

Basic stats:

```
(data['SALE PRICE'].median().item(),
data['SALE PRICE'].mean().item(),
data['SALE PRICE'].max().item())
```

```
(1255000.0, 4039845.3967075534, 357000000)
```

```
data['SALE PRICE'].describe()
```

```
count    1.239200e+04
mean     4.039845e+06
std      1.566082e+07
min      1.041490e+05
25%      6.943750e+05
50%      1.255000e+06
75%      2.675000e+06
max      3.570000e+08
Name: SALE PRICE, dtype: float64
```

Histogram (value frequency plot – does not work well due to outlier prices) – see Figure 2.2:

```
plt.hist(data['SALE PRICE'], bins=50)
plt.xlabel("Sale price")
plt.ylabel("Frequency")
plt.show()
```

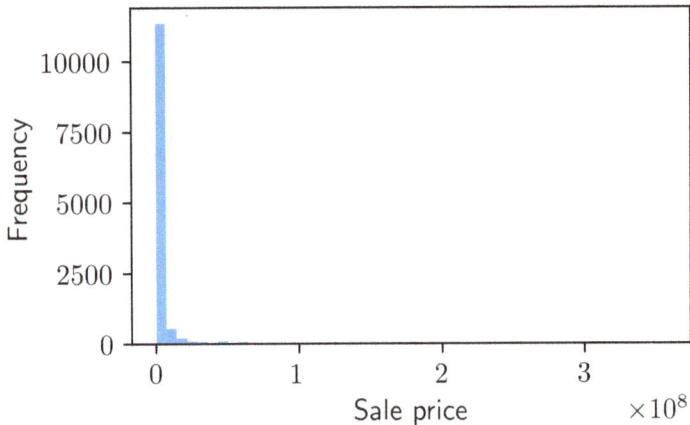

Figure 2.2: Sale price histogram

Log10 scaled price (what power to bring a 10 to in order to get a price – i.e., the order of number of zeros in a price) – simple and fancy ([Oveb]) x-axes – see Figure 2.3:

```
# regular x-axis
data['SALE PRICE LOG10'] = np.log10(data['SALE PRICE'])
plt.hist(data['SALE PRICE LOG10'], bins=50)
plt.xlabel("Log10(Sale price)")
plt.ylabel("Frequency")
plt.show()

# fancy x-axis
hist, bins = np.histogram(data['SALE PRICE'], bins=50)
lb = np.logspace(np.log10(bins[0]), np.log10(bins[-1]),
 ↪  len(bins))
plt.hist(data['SALE PRICE'], bins=lb)
plt.xscale('log')
plt.xlabel("Sale price, Log10 scale")
plt.ylabel("Frequency")
plt.show()
```

Scatter plot (sales price vs. the construction year of the property) – see Figure 2.4:

```
plt.scatter(data["YEAR BUILT"], data['SALE PRICE LOG10'],
  c = data['TAX CLASS COMMERCIAL']) # tax class color code
plt.xlabel("Year built")
plt.ylabel("Log10(Sale price)")
plt.show()
```

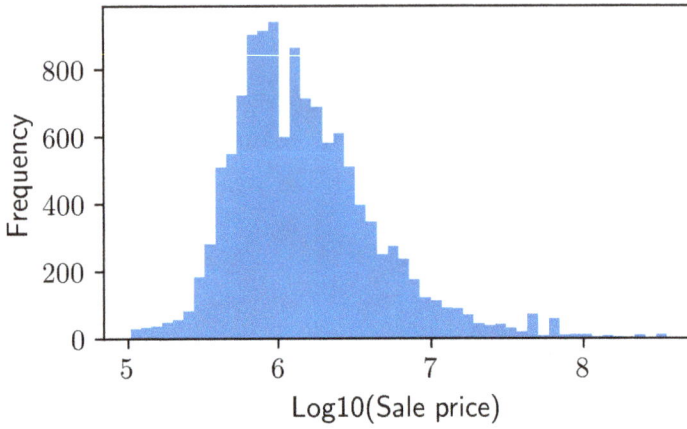

(a) Regular x-axis

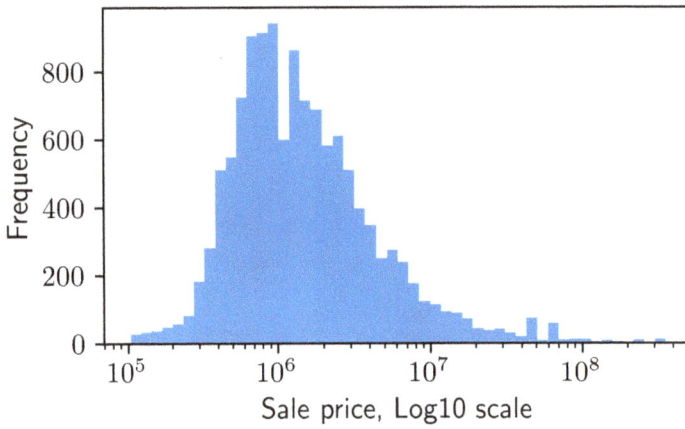

(b) Fancy x-axis

Figure 2.3: Log10 sale price histogram

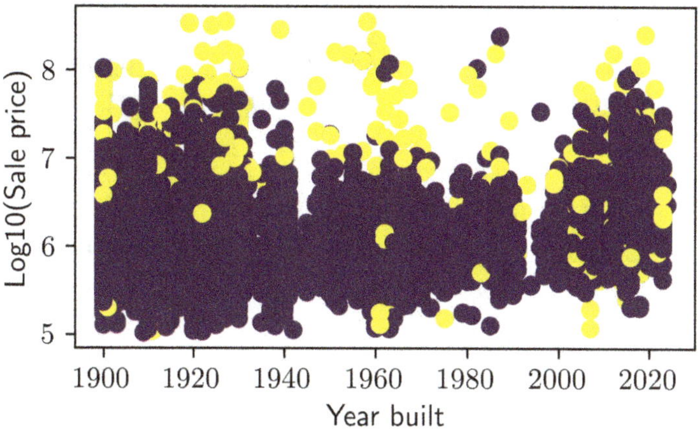

Figure 2.4: Log10 sale price by year built and tax status

Correlation:

```
data[["YEAR BUILT", 'SALE PRICE LOG10']].corr()
```

	YEAR BUILT	SALE PRICE LOG10
YEAR BUILT	1.000000	0.112292
SALE PRICE LOG10	0.112292	1.000000

groupby function:

```
data.groupby("NEIGHBORHOOD")['SALE
↪   PRICE'].median().reset_index().sort_values(
    'SALE PRICE', ascending = False)
```

	NEIGHBORHOOD	SALE PRICE
17	JAVITS CENTER	5255250.0
19	LITTLE ITALY	4540000.0
28	SOHO	3785000.0

103

	NEIGHBORHOOD	SALE PRICE
3	CIVIC CENTER	2997499.5
30	TRIBECA	2645000.0
6	FASHION	2475000.0
33	UPPER EAST SIDE (96-110)	2050000.0
11	GREENWICH VILLAGE-WEST	1870000.0
8	FLATIRON	1805000.0
2	CHINATOWN	1800000.0
22	MIDTOWN CBD	1649630.0
1	CHELSEA	1610000.0
10	GREENWICH VILLAGE-CENTRAL	1544500.0
5	EAST VILLAGE	1446924.0
35	UPPER WEST SIDE (79-96)	1402500.0
31	UPPER EAST SIDE (59-79)	1360000.0
24	MIDTOWN WEST	1332500.0
32	UPPER EAST SIDE (79-96)	1300000.0
9	GRAMERCY	1299134.5
34	UPPER WEST SIDE (59-79)	1291589.0
4	CLINTON	1245000.0
25	MORNINGSIDE HEIGHTS	1240000.0
7	FINANCIAL	1230000.0
36	UPPER WEST SIDE (96-116)	1200000.0
20	LOWER EAST SIDE	1122960.0
27	ROOSEVELT ISLAND	970000.0
26	MURRAY HILL	947500.0
21	MANHATTAN VALLEY	930000.0
29	SOUTHBRIDGE	900000.0
18	KIPS BAY	877500.0
12	HARLEM-CENTRAL	824000.0
23	MIDTOWN EAST	800000.0
14	HARLEM-UPPER	735000.0
0	ALPHABET CITY	712500.0
13	HARLEM-EAST	707000.0
37	WASHINGTON HEIGHTS LOWER	662500.0
38	WASHINGTON HEIGHTS UPPER	525000.0
16	INWOOD	472000.0

15	HARLEM-WEST	447500.0

Pivot table:

```
data.pivot_table(values="SALE PRICE", index="NEIGHBORHOOD",
↪  columns="TAX CLASS COMMERCIAL",
             aggfunc="median").sort_values(1.0,
↪  ascending=False)
```

TAX CLASS COMMERCIAL NEIGHBORHOOD	0.0	1.0
SOUTHBRIDGE	900000.0	62432448.0
MIDTOWN EAST	755000.0	60811441.0
FASHION	1731250.0	21340000.0
MIDTOWN WEST	1210000.0	18500000.0
MURRAY HILL	910000.0	15250000.0
JAVITS CENTER	5255000.0	14000000.0
HARLEM-UPPER	725000.0	12800000.0
SOHO	3050000.0	9250000.0
UPPER WEST SIDE (79-96)	1400000.0	9150000.0
GRAMERCY	1283437.5	7790486.0
FINANCIAL	1165000.0	7700000.0
GREENWICH VILLAGE-CENTRAL	1498000.0	7511252.0
GREENWICH VILLAGE-WEST	1750000.0	7500000.0
LITTLE ITALY	4237500.0	7500000.0
UPPER EAST SIDE (59-79)	1325000.0	5725000.0
MIDTOWN CBD	1475318.0	5562500.0
CHELSEA	1547500.0	5300000.0
CHINATOWN	1200000.0	4800000.0
INWOOD	457500.0	4750000.0
FLATIRON	1700000.0	4500000.0
TRIBECA	2597500.0	3790000.0
EAST VILLAGE	1380000.0	3600000.0
HARLEM-CENTRAL	799000.0	3500000.0
UPPER WEST SIDE (96-116)	1145000.0	3472600.0

TAX CLASS COMMERCIAL NEIGHBORHOOD	0.0	1.0
LOWER EAST SIDE	1040000.0	3450000.0
CIVIC CENTER	2997499.5	3400000.0
ALPHABET CITY	700000.0	2950000.0
UPPER EAST SIDE (79-96)	1275000.0	2927469.0
HARLEM-EAST	675000.0	2787500.0
WASHINGTON HEIGHTS UPPER	510000.0	2500000.0
CLINTON	1240000.0	2251826.0
KIPS BAY	830000.0	2225000.0
UPPER WEST SIDE (59-79)	1290000.0	1400000.0
WASHINGTON HEIGHTS LOWER	675000.0	500000.0
HARLEM-WEST	447500.0	NaN
MANHATTAN VALLEY	930000.0	NaN
MORNINGSIDE HEIGHTS	1240000.0	NaN
ROOSEVELT ISLAND	970000.0	NaN
UPPER EAST SIDE (96-110)	2050000.0	NaN

NYC is a pricey place.

2.6.1 Regression using Statsmodels

We will introduce the basics of linear regression in a later chapter. Roughly, this analytical technique allows us to answer how one variable changes when another variables is modified (holding other known variables constant). The code below shows how to do such analysis using Python's `statsmodels` package – we will specifically look at the expected change in the log10 price of a property (roughly, number of zeros in the price) with the changes in the year the property was constructed.

```
# !pip install statsmodels

import statsmodels.api as sm

data['Intercept'] = 1.0
```

```
X = data[['Intercept', "YEAR BUILT"]]
y = data['SALE PRICE LOG10']

model = sm.OLS(y, X).fit()
model.summary()
```

Dep. Variable:	SALE PRICE LOG10	R-squared:	0.013
Model:	OLS	Adj. R-squared:	0.013
Method:	Least Squares	F-statistic:	158.2
Date:	Tue, 09 Sep 2025	Prob (F-statistic):	4.58e-36
Time:	18:50:16	Log-Likelihood:	-8522.8
No. Observations:	12392	AIC:	1.705e+04
Df Residuals:	12390	BIC:	1.706e+04
Df Model:	1		
Covariance Type:	nonrobust		

	coef	std err	t	P> |t|	[0.025	0.975]
Intercept	3.3429	0.225	14.839	0.000	2.901	3.784
YEAR BUILT	0.0014	0.000	12.579	0.000	0.001	0.002

Omnibus:	2308.930	Durbin-Watson:	0.662
Prob(Omnibus):	0.000	Jarque-Bera (JB):	4714.193
Skew:	1.113	Prob(JB):	0.00
Kurtosis:	5.043	Cond. No.	1.02e+05

Notes:

[1] Standard Errors assume that the covariance matrix of the errors is correctly specified.

[2] The condition number is large, 1.02e+05. This might indicate that there are strong multicollinearity or other numerical problems.

```
print('Log10 price on year coefficient: ',
  ↪  round(model.params.iloc[1],3))
```

```
Log10 price on year coefficient:  0.001
```

Percent increase in price with an increase in year of construction by 1:

```
v = 100 * (10**model.params.iloc[1] - 1)
print(f"{v:.2f}%")
```

```
0.33%
```

So newer properties cost more, on average.

2.7 Discussion

In this chapter, we have seen how treating data as vectors and matrices grants us increased efficiency of both notation and computation. We have covered basic matrix algebra (dot products, transposes, etc.), how NumPy implements these ideas using array objects and broadcasting, and how Pandas builds on NumPy to offer labeled, table-based data processing. The NYC real-estate example ties the topics together, demonstrating data cleaning, analysis, and visualization. Arrays and operations on them will be indispensable to us moving forward.

2.8 Further learning resources

- Linear algebra:

    - Introduction: Chapter 2 from [JW07].
    - Intermediate: *Matrix Algebra for Engineers* course by Jeffrey Chasnov on Coursera (https://www.coursera.org/learn/matrix-algebra-engineers).
    - Advanced: https://linear.axler.net/ [Axl15].

- NumPy documentation: https://numpy.org/doc/stable/user/index.html.
- Pandas documentation: https://pandas.pydata.org/docs/.

2.9 References

[Axl15]   Sheldon Axler. *Linear Algebra Done Right*. Springer, 2015.

[BEN15]   Ray Boshara, William R Emmons, and Bryan J Noeth. "The demographics of wealth-how age, education and race separate thrivers from strugglers in today's economy. Essay No. 2: The role of education". In: *Demographics of Wealth* (2015), pp. 1–28.

[Cha25]   Jeffrey R. Chasnov. *Matrix Algebra for Engineers*. Online course on Coursera. 2025. URL: https://www.coursera.org/learn/matrix-algebra-engineers.

[Dis]   Discourse JuliaLang. *Language: "Tensor" vs. "Array"*. URL: https://discourse.julialang.org/t/language-tensor-vs-array/96590 (visited on 05/30/2025).

[Exc]   Stack Exchange. *Are one-by-one matrices equivalent to scalars?* URL: https://math.stackexchange.com/questions/65002/are-one-by-one-matrices-equivalent-to-scalars (visited on 05/02/2025).

[JW07]   Richard A. Johnson and Dean W. Wichern. *Applied Multivariate Statistical Analysis*. 6th. Prentice Hall, 2007. Chap. 2.

[OS13]   Cathy O'Neil and Rachel Schutt. *Doing Data Science: Straight Talk From the Frontline*. O'Reilly Media, Inc., 2013.

[Ovea]   Stack Overflow. *Is NumPy any faster than default Python when iterating over a list?* URL: https://stackoverflow.com/questions/73060352/is-numpy-any-faster-than-default-python-when-iterating-over-a-list (visited on 05/02/2025).

[Oveb]   Stack Overflow. *Plotting a histogram on a Log scale with Matplotlib*. URL: https://stackoverflow.com/questions/47850202/plotting-a-histogram-on-a-log-scale-with-matplotlib (visited on 05/02/2025).

[Ovec]   Stack Overflow. *Why are NumPy arrays so fast?* URL: https://stackoverflow.com/questions/8385602/why-are-numpy-arrays-so-fast (visited on 05/02/2025).

[Staa]   Stack Exchange. *An introduction to tensors*. URL: https://math.stackexchange.com/questions/10282/an-introduction-to-tensors (visited on 05/30/2025).

[Stab]   Stack Exchange. *Are there any differences between tensors and multidimensional arrays?* URL: https://math.stackexchange.com/questions/1134809/are-there-any-differences-between-tensors-and-multidimensional-arrays (visited on 05/30/2025).

[Str+07] Martinette T Streppel et al. "Mortality and life expectancy in relation to long-term cigarette, cigar and pipe smoking: The Zutphen Study". In: *Tobacco control* 16.2 (2007), pp. 107–113.

[Wika]   Wikipedia. *Basic Linear Algebra Subprograms*. URL: https://en.wikipedia.org/wiki/Basic_Linear_Algebra_Subprograms (visited on 05/02/2025).

[Wikb]   Wikipedia. *Covariance*. URL: https://en.wikipedia.org/wiki/Covariance (visited on 05/02/2025).

[Wikc]   Wikipedia. *Invertible matrix*. URL: https://en.wikipedia.org/wiki/Invertible_matrix (visited on 05/02/2025).

[Wikd]   Wikipedia. *Row- and column-major order*. URL: https://en.wikipedia.org/wiki/Row-_and_column-major_order (visited on 05/30/2025).

[Wikc]   Wikipedia. *Tensor (machine learning)*. URL: https://en.wikipedia.org/wiki/Tensor_(machine_learning) (visited on 05/30/2025).

3 Randomness and probabilities

In this chapter, we review probabilities, random numbers, and simulation in Python, which will shed light on how uncertainty can be described numerically and how deterministic algorithms can give rise to useful randomized behavior in computer programs.

3.1 What is a probability?

A **probability** is a number between 0 and 1 that quantifies how likely an event is to happen. A probability close to 1 means the event is very likely; a probability near 0 means it is unlikely.

For example, under a theoretical model of a fair coin, we expect:

- $P(\text{heads}) \approx 0.5$;
- $P(\text{tails}) \approx 0.5$;
- $P(\text{edge}) \approx 0$.

These are *theoretical* probabilities based on the physical symmetry of the coin and the assumption of an unbiased toss – not directly measured from data.

Probabilities can also be expressed in percent (%) terms – 0.5 proportion would be equivalent to 50% chance.

There are two main ways to interpret probabilities:

- **Frequentist:** Probability is the long-run relative frequency of an event in repeated, identical trials.
- **Bayesian:** Probability represents a degree of belief, informed by data, expert judgment, or other information. This notion of probability is applicable even for one-off or non-repeatable events, like a particular candidate winning a specific future political election.

For example, in the frequentist framework, we estimate the probability of heads in a coin toss as:

$$\hat{P}(\text{heads}) = \frac{\text{number of heads}}{\text{number of tosses}}.$$

Here $\hat{P}$ (with a hat) indicates an *empirical* probability, an estimate based on observed data – which could vary across different data samples.

The **Law of Large Numbers (LLN)** says that as we collect more data, empirical estimates $\hat{P}(\cdot)$ converge to the true probabilities $P(\cdot)$ – the fixed, often unknown quantities that characterize the underlying data-generating process. For example, though a short sequence of coin tosses may show a head proportion far from 0.5, as the number of coin tosses (sample size) grows, the proportion of heads will converge to ~ 0.5 – the value according to which the observed outcomes are generated by the physics of the coin toss and fall (assuming the coin is fair and the tosses are independent and representative of the process). See [Wikg; Ros13] for more details.

---

i Probability, randomness, and knowledge limitations

Probabilities reflect uncertainty and limitations in knowledge. For example, coin tosses obey Newtonian physics, so, in theory, they are perfectly deterministically predictable if we knew all the relevant physical variables. However, because the outcome of a coin toss is highly sensitive to a variety of factors that we cannot measure or control with sufficient precision (e.g., initial acceleration, air movement, slight asymmetries in the coin), we model the result as random for practical purposes – even though it is governed by deterministic physical laws.

3.2 Random variable

Here is some frequently used terminology when it comes to randomness:

- A **trial** is one instance of a random experiment – for example, a coin toss; or a throw of a six-sided die.
- An **outcome** is the result of the trial – for example, heads, tails, or coin edge; or the symbol shown on the upward side of the die.

- A **sample space** is the set of possible outcomes – for example, {heads, tails, edge}.
- A **random variable** is a variable whose value depends on the outcome. It assigns a specific numeric value to each possible outcome of a trial. For example, let X be random variable. For a coin toss, it could be defined as: $X = 1$ if the coin shows heads, $X = -1$ if tails, $X = 0$ if edge. For a throw of a die, we could define it as $X =$ number of dots on the side facing up. Before the trial, value of X is unknown. When the trial is performed, X assumes a specific value corresponding to the outcome that actually occurs. The observed value is called a *realization* of the random variable. A random variable is thus a function of an outcome – so we can, for example, equivalently define X as a mapping: $X(\text{heads}) = 1$, $X(\text{tails}) = -1$, $X(\text{edge}) = 0$.
- An **event** is a collection of outcomes, usually described in terms of the random variable – like "X is even" or "$X > 4$".
- A **random process** is a collection of random variables indexed by some parameter, for example, time. For instance, a random process X_1, X_2, X_3, ... could describe consecutive coin tosses or end-of-day stock prices on different days.

---

i Conceptual flexibility

The presented randomness-related concepts are flexible. For example, a trial could be defined to mean three consecutive coin tosses instead of a single coin toss. Then, a possible outcome would be one of the triplet sequences: "HTH", "TTT", "THH", etc. (where "T" stands for tails and "H" – for heads). A random variable X could be defined to count the number of heads in the trial. And the event of interest could be the occurrence of zero heads in a trial of three coin tosses ($X = 0$).

A **random variable** can be:

- Discrete

    - Example: rolling a six-sided die, where random variable can assume a finite discrete set of values $X \in \{1, 2, 3, 4, 5, 6\}$. (This notation means that the value of X belongs to the set $\{1, 2, 3, 4, 5, 6\}$; the symbol $\in$ means "is an element of" or "belongs to".)

- Continuous

  - Example: height in cm, where random variable can potentially assume any real number value on the interval $X \in (0, \infty)$. (Round brackets here mean edges are not included in the interval. In contrast, a square bracket next to zero $[0, \infty)$ would mean zero is included in the interval.)

- Mixed

  - Example: time until delivery in seconds, where $X = 0$ if delivery has already occurred and $X \in (0, \infty)$ otherwise.

3.3 Probability distribution

A **probability distribution** describes the likelihood of occurrence of possible events in an experiment – and thus also the likelihood of different values that the corresponding random variables may assume. There are many different ways to specify a probability distribution. Common approaches include (1) probability mass and density functions for discrete and continuous random variables respectively and (2) a cumulative distribution function – for an arbitrary random variable. Other approaches include moment generating functions (MGFs), characteristic functions (CFs), etc. – but we won't go into these more advanced topics here.

3.3.1 Discrete case: Probability mass function (PMF)

A **discrete distribution** is a probability distribution associated with a discrete random variable – a random variable that can assume a discrete set of values.

A discrete distribution can be specified by a **probability mass function (PMF)**: $P(X = x)$. In $P(X = x)$ notation, X is the discrete random variable corresponding to the distribution in question and x is the possible outcome value. For a given outcome, PMF returns a probability number. We can equivalently write $P(X = x)$ as $P_X(x)$ or just $P(x)$ for brevity – if there is no confusion about the distribution we are referring to.

For example, for a fair six-sided die, PMF is $P(X = x) = \frac{1}{6}$ for any value x that random variable X could assume from among the set $\{1, 2, 3, 4, 5, 6\}$. (This particular distribution is called a *uniform* discrete distribution because each value is *equally* likely.)

The probability of an event can then be computed as a sum of probabilities of all outcomes that satisfy the event. For instance, the probability of an event "a die rolls an even value" is just

$$P(X = 2) + P(X = 4) + P(X = 6) = 3/6 = 0.5.$$

PMF must satisfy two conditions:

- Valid probability values: $0 \leq P(X = x) \leq 1$ for all possible outcome values x.
- Total probability equals to 1: $\sum_x P(X = x) = 1$ where $\sum_x$ represents sum over all possible outcomes.

Discrete distributions over an infinite set of points are also possible. Consider an infinite sequence of probabilities defined recursively as follows:

- $p_0 = 0.5$
- $p_1 = 0.5 \cdot p_0 = 0.5^2$
- $p_2 = 0.5 \cdot p_1 = 0.5^3$
- and, in general, $p_i = 0.5^{i+1}$ for $i \geq 0$

This sequence forms a geometric series with $a = 0.5$ and ratio $r = 0.5$. By formula for the sum of the geometric series with $|r| < 1$, we have that the sum of all these probabilities is 1:[1]

$$\sum_{i=0}^{\infty} p_i = \sum_{i=0}^{\infty} ar^i = \frac{a}{1 - r} = 0.5/0.5 = 1.$$

So, in this case, as all probability values satisfy $0 \leq p_i \leq 1$ and $\sum_i p_i = 1$, this is a valid probability distribution for an infinite discrete set of points!

However, for many conceivable infinite sequences of probability-like values, their sum would not be 1 and would, in fact, go to infinity. For example, for

---

[1]*Proof:* Let $S = \sum_{i=0}^{\infty} ar^i = a + ar + ar^2 + ar^3 + ...$ Multiply both sides by r, so $r \cdot S = ar + ar^2 + ar^3 + ...$ Subtract the second equation from the first: $S - rS = a \rightarrow S = \frac{a}{1-r}$. This works only when $|r| < 1$, so the infinite geometric series converges.

$p_i = \frac{1}{i+1}$, with integer $i \geq 0$, it can be shown that $\sum_{i=0}^{\infty} p_i = \infty$ – a famous result in mathematical analysis. So such a sequence would not form a valid probability distribution.

3.3.2 Continuous case: Probability density function (PDF)

A **continuous distribution** is a distribution associated with a continuous random variable, which could, for example, assume any real number value on an interval, such as all values between 0 and 3.

Mathematically, the real numbers form an *uncountably* infinite set, which is, in a precise sense, larger than the (*countably*) infinite set of integers. Roughly speaking, this means we cannot list or index all real numbers one by one like we can with integer numbers. Whatever principle we use to enumerate real numbers (that is, put them in correspondence with non-negative integers), it is always possible to find a real number that we miss – by Cantor's diagonal argument [Wikc].

This presents a challenge when defining probabilities. If we tried to assign a non-zero probability to each individual point on the real number line (of which there are uncountably many), the total probability across all points quickly blows past 1 – violating the rule that the total probability in any distribution must equal 1. To address this issue, for continuous random variables, all individual points are assigned zero probability: $P(X = x) = 0$.

Because of this, for continuous random variables, we do not ask about the probability of an exact value – since it is always zero and thus not meaningful in practice. Instead, probabilities are defined over *intervals* of values that a continuous random variable can assume. For example, we ask questions like: *What is the probability that X lies between 1.2 and 1.5?*

One way to specify a continuous distribution is using a **probability density function (PDF)**: $p(x)$. This function does not directly return probability. Instead, it returns a real number in $[0, \infty)$ range that signals the relative likelihood of the corresponding value. However, crucially, the function $p(x)$ is defined so that the *area* under it over an interval is exactly equal to the probability that X assumes a value in that interval. So while the height of the graph of $p(x)$ is not a probability number, the area under that graph is a valid probability number.

Probabilities can then be computed over intervals via integration of the PDF $p(x)$: $P(a < X \le b) = \int_a^b p(x)dx$. The total area under $p(x)$ over the entire range of values that the random variable can take must equal to 1.

💡 Integration refresher

If you are unfamiliar with integration, think of it as summing many thin rectangles under a curve. To estimate the area under $p(x)$ between a and b, divide the interval into n slices. For each slice, multiply height at the slice's midpoint by slice's width Δx (pronounced *delta x*), and add the products up across slices. Here is an example in Python code:

```python
a, b = 1, 4
n = 1000
dx = (b - a) / n   # width of each slice

area = 0
for i in range(n):
    x_i = a + (i + 0.5) * dx    # midpoint
    area += p(x_i) * dx   # height * width
```

This summation approximates the integral:

$$\int_a^b p(x)dx \approx \sum_{i=1}^n p(x_i) \cdot \Delta x.$$

As the slice width Δx gets smaller and n gets larger, the approximation becomes more accurate. In the limit (as the number of slices goes to infinity and as $\Delta x \to 0$), this becomes the definite integral – a core tool in calculus. And if $p(x)$ is the probability density function (PDF), then the area under the curve we compute this way is exactly the probability that the continuous random variable X falls between a and b: $P(a < X \le b) = \int_a^b p(x)dx$.

For some pathological (i.e., weird) continuous random variables, the probability density function might not exist [Wikn] – for example, see Cantor distribution [Wikb].

3.3.3 Cumulative distribution function (CDF)

Alternatively, the distribution of *any* random variable can be specified by a **cumulative distribution function (CDF)**.

CDF of a random variable X gives the probability that X takes a value less than or equal to x:

$$F(x) = P(X \leq x).$$

A very useful fact is that **a CDF always exists for any random variable – discrete, continuous, or mixed**. This holds even in pathological cases when a continuous random variable PDF does not exist.

If X is discrete, continuing with our fair die example:

$$F(3) = P(X \leq 3) = P(1) + P(2) + P(3) = 1/2.$$

So, in this discrete case, the CDF at value x is just the sum of PMF values up to and including x.

If X is continuous and its PDF exists, the CDF is the integral of the PDF:

$$F(x) = \int_{-\infty}^{x} p(t)dt.$$

This gives the total probability mass accumulated up to x. (The immediate conclusion that follows from this is that a PDF function at point x is actually just a derivative of the CDF, assuming CDF is differentiable at that point. We will discuss derivatives in later chapters.)

Conveniently, CDFs give us a way to compute the probability that a random variable X falls between two values a and b:

$$P(a < X \leq b) = P(X \leq b) - P(X \leq a) = F(b) - F(a).$$

The CDF also exists for and describes *mixed random variables* – continuous random variables that include point masses, or specific values with non-zero probability.

Inverse CDF function $F^{-1}(p)$, also called a quantile function, given probability p, returns the value x, for which the CDF yields that probability:

$$F(x) = p \Rightarrow x = F^{-1}(p).$$

Example:

- If $F(3) = 0.5$, then $F^{-1}(0.5) = 3$.
- This means the median of X is 3 (50% of values are ≤ 3).

Inverse CDF is useful for random number generation, as we will see later.

---

i $\leq$ vs. $<$ in $P(X \leq x)$

In *discrete and mixed distributions*, the distinction between $\leq$ and $<$ in $P(X \leq x)$ matters: exact values x can have non-zero probability. However, in *continuous distributions*, $P(X = x) = 0$, so $P(X \leq x) = P(X < x)$ and $\leq$ and $<$ can be used interchangeably – although it is better to be consistent stylistically.

---

i Continuous distributions and limited floating-point precision

In truth, all distributions are effectively discrete on a computer – due to limited floating-point precision. Even continuous variables, like height, are represented using a finite number of values and can be treated as discrete when rounded. This approximation is typically sufficient for practical purposes. However, academic treatments usually maintain a strict distinction between discrete and continuous variables, as it leads to cleaner theory and useful mathematical results. These nuances are studied more formally in *measure theory*.

3.4 Example: Normal distribution

The **normal distribution** (bell curve) is a continuous random variable distribution widely used to model natural phenomena like adult human height. It has two parameters that fully determine its shape:

- Mean μ (center; the character is pronounced as *mu*).

- Standard deviation σ (spread; the character is pronounced as *sigma*) [σ^2 is called variance].

A normal distribution characterizes events where outcomes are more frequent the closer they are to the mean and more rare the further away they are from the mean, forming a bell-shaped curve.

The formula for a normal probability density function (PDF) is:

$$p(x) = \frac{1}{\sigma\sqrt{2\pi}} \exp\left(-\frac{(x-\mu)^2}{2\sigma^2}\right),$$

where π is the pi constant $3.14159\ldots$. Assume the adult human height follows a normal distribution with mean $\mu = 170$ cm and standard deviation $\sigma = 10$ cm. This means average-height individuals are common, and extremely short / tall people are rare, symmetrically scattered around the mean.

We can evaluate the normal density at a point using Python:

```
from scipy.stats import norm
x = 170
pdf_value = norm.pdf(x, loc=170, scale=10)
pdf_value
```

```
np.float64(0.03989422804014327)
```

`pdf_value` gives us **density** at 170 cm, **not** the probability of being exactly 170 cm tall (which is zero). Density captures how *relatively* likely that point is.

To compute a probability, we would need to compute the area under the density over desired interval – that is, the integral over the interval: $P(165 < X \le 175) = \int_{165}^{175} p(x)dx$.

This can be approximated by summation, as shown earlier:

```
a, b = 165, 175
n = 10000
dx = (b - a) / n   # width of each slice
prob = 0
```

```
for i in range(n):
    x_i = a + (i + 0.5) * dx
    prob += norm.pdf(x_i, loc=170, scale=10) * dx

print("P(165 < X <= 175) =", round(prob,3))
```

```
P(165 < X <= 175) = 0.383
```

This tells us around 38% of individuals are between 165 and 175 cm tall.

Let us also visualize the normal PDF:

```
import numpy as np
import matplotlib.pyplot as plt


mu, sigma = 170, 10
x = np.linspace(130, 210, 500)
y = norm.pdf(x, loc=mu, scale=sigma)

plt.plot(x, y, label="Normal PDF")
x_fill = np.linspace(165, 175, 300)
y_fill = norm.pdf(x_fill, loc=mu, scale=sigma)
plt.fill_between(x_fill, y_fill, alpha=0.3, color='skyblue',
 ↪  label="$P(165 < X \\leq 175)$")
plt.xlabel("Height (cm)")
plt.ylabel("Density")
plt.ylim(0,0.045)
plt.legend(loc='upper right', prop={'size': 8})
plt.grid(True, linestyle=':', linewidth=0.5)
plt.show()
plt.close()
```

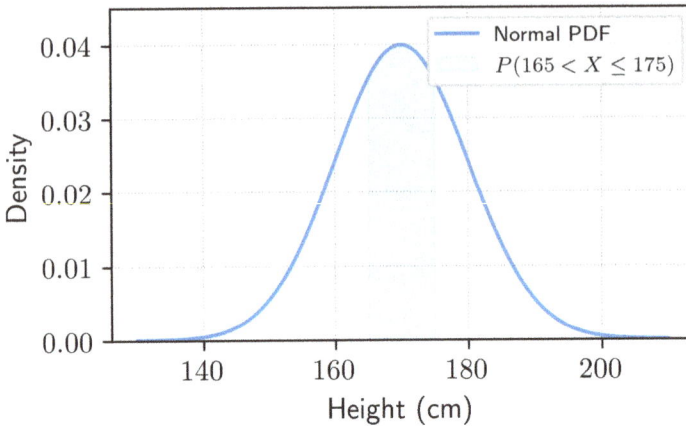

Figure 3.1: Normal PDF of heights

The shaded region in Figure 3.1 represents the area under the curve between 165 and 175 cm – which equals the probability of falling in that range.

3.5 Measures of distribution shape

Distribution moment is a name for a quantitative measure of the shape of a distribution graph. A couple of measures are particularly important when it comes to measuring the shape of a distribution.

The **first moment** of a random variable is known as the (theoretical) **mean** or the **expectation** and is one of the possible measures of a center of a distribution. It is a probability-weighted sum of possible outcomes under the distribution:

- Discrete case: $E[X] = \sum_i x_i \cdot P(X = x_i)$.
- Continuous case: $E[X] = \int_{-\infty}^{\infty} x \cdot p(x)dx$.

Empirical (sample) mean is an arithmetic average of observed random variable values in a data sample (e.g., actual sequence of coin tosses):

$$\bar{x} = \frac{1}{n} \sum_{i=1}^{n} x_i.$$

By the Law of Large Numbers (LLN), the empirical mean converges to the expectation $\bar{x} \to E[X]$ as the data sample grows large $n \to \infty$ (if samples are independent identically distributed (i.i.d.)). See [Wikg; Ros13].

The **second moment** (around the mean) is called **variance**. Consider subtracting from a random variable X its theoretical mean $E[X]$ and taking the square of the difference: $(X - E[X])^2$. This is, in fact, a new random variable – that depends on realization X. Now, we can take its expectation – the probability-weighted sum of all possible outcomes of this new transformed variable of squared deviations from the mean. This measure is called variance:

- Discrete case:

    - $\sigma^2 = \mathrm{Var}(X) = E[(X - E[X])^2] = \sum_i (x_i - E[X])^2 \cdot P(X = x_i)$.

- Continuous case:

    - $\sigma^2 = \mathrm{Var}(X) = E[(X - E[X])^2] = \int_{-\infty}^{\infty} (x - E[X])^2 \cdot p(x)dx$.

We can also expand $\mathrm{Var}(X) = E[(X - E[X])^2] = E[X^2] - (E[X])^2$ due to how probability-weighting distributes.

The **standard deviation** is the square root of the variance: $\sigma = \sqrt{\mathrm{Var}(X)}$.

Variance captures how widely values are spread around the mean.

Empirical (sample) variance is an arithmetic average of squared deviations $s^2 = \frac{1}{n} \sum_{i=1}^{n} (x_i - E[X])^2$. If we do not know the theoretical first moment $E[X]$, but instead use sample mean $\bar{x}$ in its place, we need to adjust the formula to $s^2 = \frac{1}{n-1} \sum_{i=1}^{n} (x_i - \bar{x})^2$. This is Bessel's correction – it ensures that sample variance correctly goes to the expected variance $s^2 \to \sigma^2$ as $n \to \infty$ via LLN.

Other measures – moments of higher order – exist, such as skewness and kurtosis – but these are more rarely used. See [Wikk; BH19].

3.6 Summary: Uniform and normal distributions

We will summarize here the data on the uniform and normal distributions. Other distributions like the binomial, Poisson, and exponential are also common in statistics and machine learning. See [Wiki] for an extensive list of known distributions.

3.6.1 Uniform distribution

Discrete uniform

Each of k possible outcomes has equal probability.

- Example: fair six-sided die $\Rightarrow P(X = x) = 1/6$ for $x \in \{1, 2, 3, 4, 5, 6\}$.
- Probability mass function: $P(X = x) = 1/k$.

Continuous uniform

- Probability density is constant on interval $[a, b]$:

$$p(x) = \begin{cases} \frac{1}{b-a} & \text{for } a \leq x \leq b, \\ 0 & \text{otherwise.} \end{cases}$$

  This is just a horizontal line. That is, uniform probability density is the same for every x in the specified range.

- Mean: $E[X] = (a + b)/2$.
- CDF of a continuous uniform distribution is $F(x) = P(X \leq x) = \int_a^x \frac{1}{b-a} dt = \frac{1}{b-a} \int_a^x 1 dt = \frac{x-a}{b-a}$.

3.6.2 Normal distribution

We already introduced the (univariate) normal distribution in detail earlier. In brief:

- Alternative names: Gaussian, bell curve.
- Shape: Bell-shaped, symmetric around the mean.
- Fully specified by mean $E[X] = \mu$ and standard deviation $\sigma = \sqrt{\text{Var}(X)}$.
- PDF:

$$p(x) = \frac{1}{\sigma\sqrt{2\pi}} \exp\left(-\frac{(x-\mu)^2}{2\sigma^2}\right).$$

- **Standard** normal distribution is a normal distribution with mean 0 and standard deviation 1, commonly written as Normal$(0, 1)$.

- $\pm\sigma$ standard deviation interval around the mean contains 68.27% of probability mass under normal PDF; $\pm 2\sigma$ contains 95.45%; and $\pm 3\sigma$ contains 99.73%.
- It can be shown that a linear combination (weighted sum) of two independent normal random variables is also normally distributed. Let independent $z_1 \sim N(\mu_1, \sigma_1^2)$ and $z_2 \sim N(\mu_2, \sigma_2^2)$. Then, for any constants w_1 and w_2, the linear combination $w_1 z_1 + w_2 z_2$ follows normal distribution: $w_1 z_1 + w_2 z_2 \sim N(w_1 \mu_1 + w_2 \mu_2, w_1^2 \sigma_1^2 + w_2^2 \sigma_2^2)$. Independence is sufficient but not necessary – if z_1 and z_2 are jointly normal (even if dependent), then

$$w_1 z_1 + w_2 z_2 \sim N(w_1 \mu_1 + w_2 \mu_2, w_1^2 \sigma_1^2 + w_2^2 \sigma_2^2 + 2 w_1 w_2 \mathrm{Cov}(z_1, z_2)).$$

One can prove this via a **PDF** convolution [Wikf; Ros13] or using moment generating functions (MGFs) [Wikl; Ros13]; these are advanced topics that we do not cover here.

Normal density can be viewed as a sequence of transformations on x – for example, for $\mu = 2$, $\sigma = 2$, see Figure 3.2 (this example is based on a lecture slide [Orb]).

Thus, mechanistically, a normal density is an exponentiated negative quadratic function (upside-down u-shaped parabola) $\exp(-x^2)$, but appropriately scaled and shifted.

The normal distribution is critical because it can be shown, for example, that the **sum** of **any** n random variables drawn independently from some distribution with finite mean and variance will be distributed more and more as the normal distribution as n grows large – even if the random variables themselves are not normally distributed. This result is known as the **Central Limit Theorem (CLT)**. See [Wikd; Ros13].

For example, it turns out that adult human height is distributed roughly normal – perhaps, because it is a result of summation of multiple small random influences from our genes and environment.

Importantly, the arithmetic mean of random variables in a finite sample is just their normalized sum – and so is also Normally distributed as the sample size grows large – around the true expectation of the source distribution.

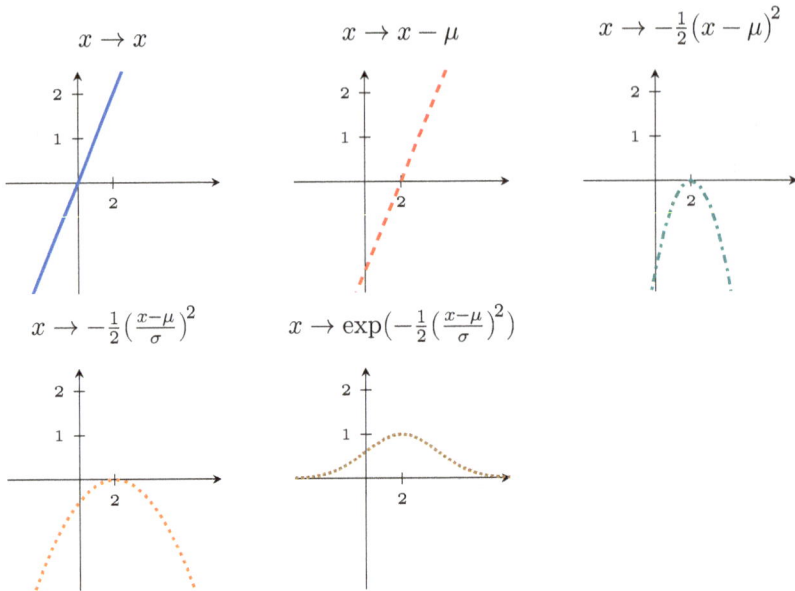

$$x \to x \qquad x \to x - \mu \qquad x \to -\tfrac{1}{2}(x - \mu)^2$$

$$x \to -\tfrac{1}{2}\left(\tfrac{x-\mu}{\sigma}\right)^2 \qquad x \to \exp\left(-\tfrac{1}{2}\left(\tfrac{x-\mu}{\sigma}\right)^2\right)$$

Figure 3.2: Normal PDF components ($\mu = 2$, $\sigma = 2$)

3.6.3 Multivariate normal distribution

The multivariate normal distribution generalizes the normal distribution to multiple variables. Instead of modeling a single variable like height or weight, it models a vector of continuous random variables that may be statistically correlated.

The multivariate normal distribution is denoted as

- $X \sim N(\mu, \Sigma)$,

where:

- X is a vector of k random variables, e.g., $(X_1, X_2, ..., X_k)$;
- μ is a vector of k means: $\mu = E[X]$;
- Σ is a $k \times k$ covariance matrix (as covered in the previous chapter): $\Sigma_{ij} = \text{Cov}(X_i, X_j)$.

Example: Height and weight in a population can often be modeled as jointly normally distributed – individuals who are taller tend to weigh more, and the joint distribution captures both their individual variation and their positive correlation.

If the variables are independently normally distributed, all values off the main diagonal in the covariance matrix Σ are zero. Drawing samples from such a multivariate distribution is equivalent to drawing samples from individual univariate normal distributions.

3.7 Conditional probability

A **conditional probability** is the probability of one event, given that another has occurred.

For two events A and B, probability of A occurring given (conditional on) B occurring is written as $P(A \mid B)$.

Example:

- A: person is taller than 180 cm
- B: person is a woman

Then $P(A \mid B) = P(> 180 \text{ cm height} \mid \text{woman})$ is the probability that a person is over 180 cm given that the person is a woman. In the USA, we could expect that $P(A \mid B) < P(A)$ – because conditioning on someone being a woman reduces the likelihood of them being taller than 180 cm compared to a randomly selected person from the general USA population.

Conditional probability of A given B – $P(A \mid B)$ – is computed, by definition, as division of the probability of two events A and B happening at the same time – $P(A \cap B)$ – by the probability of the event B we are conditioning on – $P(B)$:

$$P(A \mid B) = \frac{P(A \cap B)}{P(B)} \quad \text{for } P(B) > 0.$$

This can be estimated from data using relative frequencies:

$$\hat{P}(A \mid B) = \frac{\hat{P}(A \cap B)}{\hat{P}(B)}$$

$$= \frac{\text{number of women} > 180 \text{ cm}/\text{number of people}}{\text{number of women}/\text{number of people}}$$

$$= \frac{\text{number of women} > 180 \text{ cm}}{\text{number of women}}.$$

This tells us: among all women, what fraction are taller than 180 cm?

Bayes theorem. We can rewrite $P(A \mid B) = \frac{P(A \cap B)}{P(B)}$ as $P(A \cap B) = P(A \mid B) \cdot P(B)$ – by multiplying by $P(B)$ on both sides (if it is not zero). In the context of our example, the proportion of tall women in the population is the product of (1) the proportion of tall women among all women and (2) the proportion of women in the population.

Further, by symmetry, $P(A \cap B) = P(A \mid B) \cdot P(B) = P(B \mid A) \cdot P(A)$. That is, we can alternatively compute the proportion of tall women in the population as the product of the proportion of women in tall human segment and the proportion of the tall human segment in the overall population.

This identity can be re-arranged into the formula for conditional probability known as the Bayes theorem:

$$P(A \mid B) = \frac{P(B \mid A) \cdot P(A)}{P(B)}.$$

This is useful when it is easier to reason about $P(B \mid A)$ (e.g., probability of a positive test given a disease – we can relatively easily measure this among verifiably sick individuals), and we want to know $P(A \mid B)$ (probability of disease given a positive test – which could be costly to measure directly, e.g., for rare diseases).

Bayes theorem is central to many applications, including:

- Medical diagnostics;
- Spam filtering;
- Probabilistic inference in machine learning.

Do not be intimidated by the formula. It is just algebra applied to table counts.

---

i Law of total probability

Sometimes, we also expand the denominator as $P(B) = P(B|A) \cdot P(A) + P(B|\text{not } A) \cdot P(\text{not } A)$. This expansion just says: to get the total probability of B, we add up the probabilities of B happening in each possible scenario (whether A happens or not), weighed by how likely each scenario is.

Independence. If knowing B gives no information about A, we say A and B are independent: $P(A \mid B) = P(A)$. For independent random variables, we get $P(A \cap B) = P(A \mid B) \cdot P(B) = P(A) \cdot P(B)$.

💡 Chain rule of probability

The chain rule allows us to break down the joint probability of multiple events into a product of conditional probabilities. For three events A, B, and C, we can write: $P(A \cap B \cap C) = P(A|B \cap C) \cdot P(B \cap C) = P(A|B \cap C) \cdot P(B \mid C) \cdot P(C)$. This expresses the joint probability as the probability of C, times the probability of B given C, times the probability of A given both B and C. This generalizes to more variables, allowing us to write complex joint probabilities as a product of sequential conditional terms. **If** events for A, B, C were **mutually independent**, then all conditional probabilities become unconditional, so $P(A \cap B \cap C) = P(A) \cdot P(B) \cdot P(C)$.

While we have illustrated conditional probability using binary events, the concept generalizes naturally to multi-level categorical variables and continuous random variables. For example, suppose we are interested in $P(\text{occupation} = \text{engineer} \mid \text{education} = \text{master's})$. This is a valid conditional probability between non-binary variables. In this case, we would compute the conditional distribution of one variable given the level of another – such as the full distribution of occupations among individuals with a master's degree. This is especially useful in social science, survey analysis, and classification problems.

For continuous variables, conditional probability is defined using **condi-**

tional probability density functions. For X and Y continuous:

$$p(y \mid x) = \frac{p(x, y)}{p(x)}, \quad \text{with } p(x) > 0.$$

Here, $p(y \mid x)$ is the conditional density of Y given $X = x$. This expression is conceptually similar to the discrete case, but involves densities and calculus rather than probabilities and sums. For example, $p(\text{weight} \mid \text{height} = 180)$ tells us how body weights are distributed among people of height 180 cm. See more on this here: [Wike; Ros13]. (As a technical note, although the event height $= 180$ has zero probability in a continuous distribution, conditioning on this event is still meaningful because we are working with density functions, not probabilities.)

Using the conditional probability concept, we can compute conditional moments, such as the *conditional expectation* – the expected value of a random variable given some known information or condition. For example, $E[\text{weight} \mid \text{height} = 180] = \int_{-\infty}^{\infty} \text{weight} \cdot p(\text{weight} \mid \text{height} = 180) dweight$ gives the expected weight among people who are 180 cm tall.

3.8 Random number generation

It is often necessary to generate random numbers using a computer. For example, random numbers can be used to subsample or shuffle data, sample a next word in a large language model, simulate economic conditions and financial returns, generate environment in games, or obtain a random key for data encryption.

Where do we get the random numbers from?

In the *physical world*, we can source randomness from unpredictable processes like radioactive decay, thermal noise (electronic noise generated by the thermal agitation of the charge carriers), or even patterns in lava lamps. These hardware-based sources of randomness are useful in contexts where unpredictability is crucial (e.g. cryptographic applications).

In data science applications, however, it is much more critical to have a sequence of random numbers that one can reproduce (replicate) at will. For example, you might run a financial simulation and want someone else to be able to reproduce your exact results. That is where pseudo-random number generators (PRNGs) come in.

3.8.1 Pseudorandom number generators

A pseudorandom number generator (PRNG) is an algorithm that uses a deterministic procedure and a seed (initial) value to generate a sequence of numbers that appear random. Given the same seed, a PRNG will always produce the same sequence – perfect for reproducibility in scientific computing.

One of the simplest PRNGs is the Linear Congruential Generator (LCG), defined by:

$$X_{n+1} = (aX_n + c) \bmod m$$
$$U_n = X_n/m,$$

where X_0 is the initial seed number, $0 < a < m$ and $0 \leq c < m$ are constant integers, **mod** is a modulo division operator (% in Python; returns the remainder of a division – e.g., 6 % 4 = 2 , 4 % 4 = 0 , 2 % 4 = 2). U_n is the resulting nth pseudorandom number, which, due to normalization by m, lies between 0 (included) and 1 (not included), i.e., in the $[0, 1)$ interval. If the parameters are chosen carefully (see Hull-Dobell theorem [Wikh]), an LCG can cycle through all values in its range before repeating – this is called a full-period generator.

Simple LCG in Python, using values for LCG constants from [Fla+92]:

```python
def lcg(seed, a, c, m, n):
    u = []
    x = seed
    for _ in range(n):
        x = (a * x + c) % m    # % is modulo operator
        u.append(x / m)    # normalizing to [0, 1)
    return u

seed = 42
a = 1664525
c = 1013904223
m = 2**32
```

```
n = 10

lcg(seed, a, c, m, n)


[0.2523451747838408,
 0.08812504541128874,
 0.5772811982315034,
 0.22255426598712802,
 0.37566019711084664,
 0.02566390484571457,
 0.4472812858875841,
 0.1184600037522614,
 0.8738137057516724,
 0.9946342753246427]
```

The resulting values behave similarly to samples from a continuous Uniform$(0, 1)$ distribution (if $U \sim$ Uniform$(0, 1)$, U is equally likely to assume any value on 0-1 interval).

> **! Better PRNGs**
>
> LCGs are fast and easy to implement, but they have limitations, especially for high-dimensional simulations, and should not be used for cryptographic purposes. Modern libraries instead rely on more sophisticated PRNGs such as:
>
> - Mersenne Twister (`random` module in Python) [Wikj]
> - PCG-64 (`numpy.random.default_rng()` in NumPy) [Num]
>
> A good PRNG ensures the resulting samples comply with all theoretical properties of samples from the Uniform$(0, 1)$ distribution, as measured by statistical tests.

It is actually quite a striking fact of nature that a fully deterministic, replicable, and transparent procedure – like a PRNG – can generate a sequence of numbers that passes nearly all statistical tests for randomness. Then, just because a sequence appears unpredictable or chaotic does not mean it was generated by a truly random process. Some deterministic systems can produce outcomes that are *indistinguishable* from randomness to an outside observer. This insight is crucial in fields like cryptography, chaos theory, and even physics, where what seems like randomness may stem from the underlying unobserved order.

3.9 Sampling from (almost) any distribution

Once we can generate a stream of uniform random numbers $U \sim$ Uniform$(0, 1)$ using a PRNG, we can use a mathematical trick to generate random numbers from any other distribution.

Recall that CDF of a random variable X gives the probability that X takes a value less than or equal to x: $F(x) = P(X \leq x)$. Inverse CDF $F^{-1}(p)$ (quantile function) takes as input a probability p and returns the value x, for which the CDF yields that probability: $F(x) = p \Rightarrow x = F^{-1}(p)$.

It turns out that if U is sampled from Uniform$(0, 1)$, then $X = F^{-1}(U)$ is a sample from the desired target distribution that is described by the CDF $F(x)$!

For example, using the inverse CDF of the normal distribution (implemented as `scipy.stats.norm.ppf()`), we can generate standard normal random variables from uniform random numbers. (Normal CDF and inverse CDF functions do not have nice closed form expressions, but are implemented in all common software.)

This method is called **inverse transform sampling** and is foundational in simulation and probabilistic modeling.

Example of sampling from standard normal (mean $\mu = 0$ and standard deviation $\sigma = 1$) via transformation of uniform random draws:

```python
from scipy.stats import norm
import numpy as np

np.random.seed(42)
u = np.random.uniform(0, 1, 5)  # 5 draws from Uniform(0,1)
x = norm.ppf(u)                 # convert to Normal(0,1)
print(x)
```

```
[-0.31985238  1.65181933  0.61885465  0.24987627 -1.01095644]
```

We can draw samples from the multivariate normal distribution quite simply too. To get a random vector X of length k, $X \sim N(\mu, \Sigma)$, first draw a vector Z as k independent standard normal random variables (mean 0, standard deviation 1) – e.g., via the code snippet above. Next, compute the *Cholesky decomposition* (square root) of the covariance matrix: $\Sigma = LL^T$ (`numpy.linalg.cholesky()` function in Python). Then set $X = \mu + LZ$, where LZ is a dot product of L and Z. The resulting vector X will be distributed $N(\mu, \Sigma)$.

If we want to simulate a discrete distribution, such as a coin flip, this is also straightforward using uniform random numbers and thresholding. For example:

```
U = np.random.uniform(0, 1)
flip = 1 * (U > 0.5)
```

This simple logic returns 1 if the uniform random number exceeds 0.5, and 0 otherwise – producing a fair 50 / 50 coin flip. This method generalizes: for any discrete distribution, we can use thresholds based on cumulative probabilities to assign values. For example, if we want to sample from a 3-outcome discrete distribution with probabilities `[0.2, 0.5, 0.3]` , we can divide the [0,1] interval into segments of lengths 0.2, 0.5, and 0.3, and assign the outcome based on where a uniform draw falls.

Here is an example of how we could sample k integers uniformly *with replacement* (meaning, the same integer value may be selected multiple times), from a $[0; n)$ interval using uniform random draws:

```
# sampling integers WITH replacement

np.random.seed(42)

n = 10  # integer range: 0 to n-1 (edges included)
k = 20  # sample size
```

```
# get k uniform random variables
u = np.random.uniform(0, 1, size=k)

# multiply by n and round to the lower integer
samples = np.floor(n * u).astype(int)

print(samples)
```

```
[3 9 7 5 1 1 0 8 6 7 0 9 8 2 1 1 3 5 4 2]
```

To sample *k* *unique* integers (i.e., *without replacement*) uniformly from a $[0; n)$ interval, we could use a *shuffle* based on uniform random variables:

```
# sampling integers WITHOUT replacement

np.random.seed(42)

n = 10   # integer range: 0 to n-1 (edges included)
k = 5    # sample size

# get n uniform random variables
# (a value for each possible integer)
u = np.random.uniform(0, 1, size=n)

# get array of all possible unique integers
a = np.arange(n)

# random shuffle
# sort integers based on their associated uniform random
   ↪   values
a_sorted = a[np.argsort(u)]

# take the first k integers — sample without replacement
samples = a_sorted[:k]

print(samples)
```

```
[6 5 4 0 3]
```

These code snippets could be used to select a subset of rows in a data set.

And so we have a recipe to generate random numbers from any distribution that has a computable inverse CDF – which is sufficient for our purposes. Alternative approaches to generating non-uniform random numbers include acceptance-rejection and convolution methods for continuous random variables and alias method for discrete random variables [Wikm].

3.10 Simulation exercise

The general term "simulation" means imitation. However, we will use this term to refer specifically to **Monte Carlo simulation** – a procedure involving repeated random sampling to mimic a real or theoretical process and estimate the probability of the outcomes of interest. We will now consider several simulation examples to practice the theoretical concepts we have covered so far.

3.10.1 Adult human height distribution

Let us start by generating random numbers from the **standard normal distribution** (mean 0, standard deviation 1). We use a fixed random seed for reproducibility.

```
np.random.seed(999)
np.random.randn(4,2)  # 4 x 2 matrix of standard normal
 ↪  random variables
```

```
array([[ 0.12715784,  1.40189088],
       [ 0.31481499, -0.85844916],
       [-0.26613444, -0.64890071],
       [ 1.56626757, -2.09137019]])
```

To sample from a normal distribution with arbitrary mean and standard deviation, we use:

```
np.random.seed(999)
mu = 0
sd = 1
np.random.normal(mu, sd, (4,2))
```

```
array([[ 0.12715784,  1.40189088],
       [ 0.31481499, -0.85844916],
       [-0.26613444, -0.64890071],
       [ 1.56626757, -2.09137019]])
```

As introduced earlier, adult height tends to follow a normal distribution. Knowing the mean and standard deviation of height allows us to model the full distribution and estimate probabilities such as the chance of being above or below a certain threshold.

Let us work through a simple hypothetical example.

Part A:

- Population male adult height in cm follows a normal distribution:
  - Mean: $\mu_{male} = 175$
  - Standard deviation: $\sigma = 8$
  - $X \sim \text{Normal}(\mu = 175 \text{ cm}, \sigma = 8 \text{ cm})$

- What is the probability that a randomly selected man is taller than 185 cm?
  - $P(\text{height} > 185 \mid \text{gender} = \text{male})$

Sample height for 1 million men:

```
mu_male = 175
sd = 8
```

```
np.random.seed(999)
```

```
heights_men = np.random.normal(mu_male, sd, 1000000)
```

Estimate the probability using the relative frequency:

```
np.sum(heights_men >= 185)/heights_men.shape[0]
```

```
np.float64(0.104927)
```

Same as taking a mean on a True / False binary indicator vector:

```
np.mean(heights_men >= 185)
```

```
np.float64(0.104927)
```

This works because Boolean comparisons return True / False, which convert to 1 / 0 in numerical context – so `np.mean` computes the proportion.

Let us visualize the distribution:

```
plt.hist(heights_men, bins=100)
plt.xlabel("Height (cm)")
plt.ylabel("Frequency")
plt.axvline(x=185, color='r', label="Threshold (185 cm)")
plt.legend(prop={'size': 8})
plt.show()
```

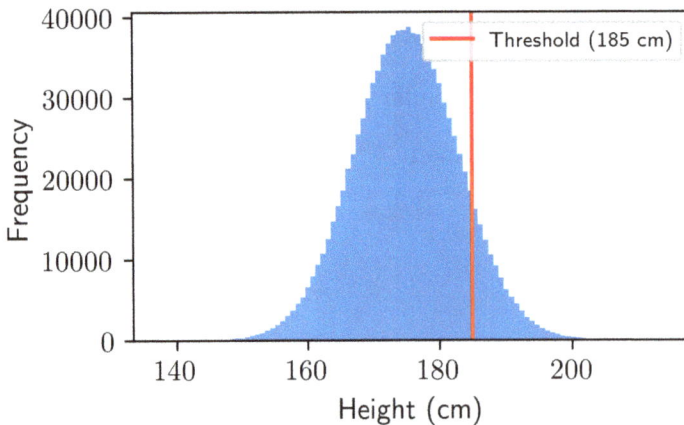

Figure 3.3: Histogram of male heights

Part B:

Let us consider a reverse question involving two genders.

- What is the probability that a random person taller than 185 cm is a woman?

    - $P(\text{gender} = \text{female} \mid \text{height} > 185)$
    - $\mu_{\text{female}} = 161$
    - Standard deviation is the same as male
    - Population split: 50% men, 50% women

Simulate and estimate:

```
mu_female = 161
heights_women = np.random.normal(mu_female, sd, 1000000)
heights = np.concatenate([heights_women, heights_men])
gender = np.concatenate([np.ones(1000000),
  ↪  np.zeros(1000000)]) # 1==female, 0==male
gender[heights > 185].mean()
```

```
np.float64(0.012814119993602349)
```

This is an empirical estimate of $P(\text{female} \mid \text{height} > 185)$, based on simulated data. It illustrates how conditional probabilities can be estimated using subset filtering.

3.10.2 Central Limit Theorem and random variable sums

Let us draw 1,000 samples, each containing 100 values drawn from a uniform distribution over the interval $[0, 1]$. We will then compute the mean of each sample:

```
np.random.seed(999)
U = np.random.uniform(0, 1, (1000, 100))
X = np.mean(U,1)
print(X.shape)
```

```
(1000,)
```

Now let us visualize the distribution of these sample means using a frequency histogram:

```
plt.hist(X, bins=30)
plt.xlabel("Average value")
plt.ylabel("Frequency")
plt.title("Histogram of sample means")
plt.show()
```

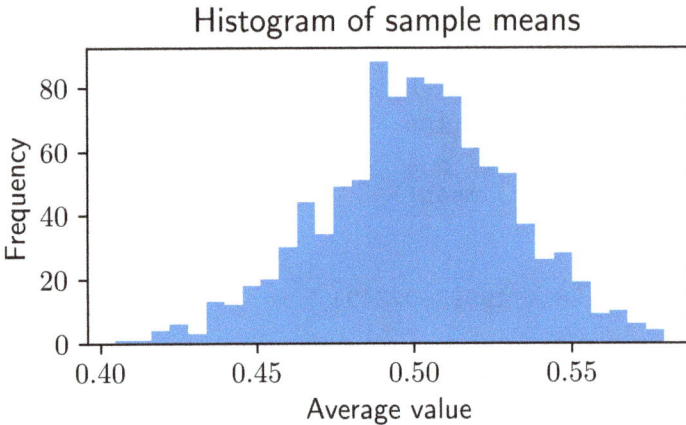

Figure 3.4: Histogram of sample means from uniform distribution

The resulting plot already resembles a normal distribution.

To make this more apparent, we can plot the empirical density (normalizing the histogram area to equal 1) and overlay the corresponding normal probability density function (PDF) using the sample mean and standard deviation:

```
# relative frequency histogram
plt.hist(X, bins=30, density=True, label="Empirical density")
```

```
# normal PDF for comparison
x = np.linspace(X.min(), X.max(), 500)
y = norm.pdf(x, loc=X.mean(), scale=X.std())
plt.plot(x, y, label="Normal PDF")

plt.xlabel("Average value")
plt.ylabel("Density")
plt.title("Empirical vs. normal distribution")
plt.legend(loc='upper right', prop={'size': 8})
plt.grid(True, linestyle=':', linewidth=0.5)
plt.show()
```

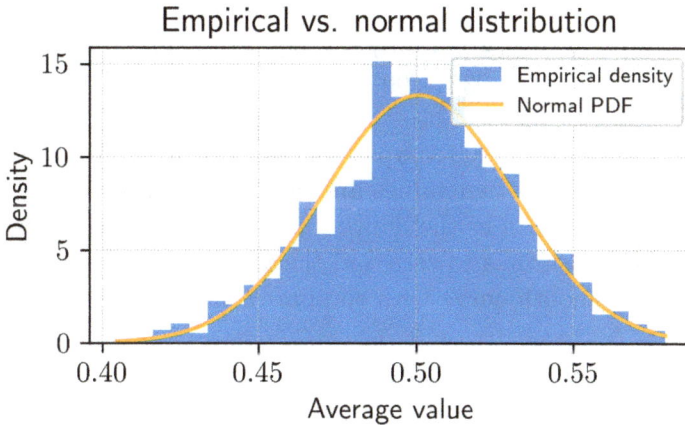

Figure 3.5: Empirical density of sample means from uniform distribution vs. theoretical normal density

As you can see in Figure 3.5, the normal distribution closely matches the distribution of averages of samples drawn from the uniform distribution.

i Density is not probability

The histogram is normalized to show probability density, not probability. When `density=True` in `plt.hist()` , we are telling Matplotlib: "Scale the bar heights so that the total area of the histogram sums to 1." This way, the area of each bar (not its height) represents the probability of a value falling into the corresponding bin. The area of the bar is calculated as (bin width × bar height). So the height of each bar – that is, the empirical density – is the relative frequency (probability) of observation belonging to a bin divided by the bin width. If the bin width is small – as it is here with 30 bins, and a lot of the probability mass is concentrated in a narrow range – as is the case with sample means due to the Central Limit Theorem, then density values can easily exceed 1, especially near the center of the distribution.

For example, if a histogram has 30 bins and the spread of the sample means is very tight (say, from 0.45 to 0.55), then:

- Each bin width is approximately $0.1/30 \approx 0.0033$.

- If $\sim 30\%$ of values fall into one bin, the height of that bin will be around $0.3/0.0033 \approx 90$.

That is why even with 1000 samples, densities of 10 or more are totally plausible, especially when values are tightly concentrated. In fact, a PDF can be thought of as the limiting form of a normalized histogram as the bin width shrinks to zero and the number of samples increases – producing a smooth curve with total area under it equal to 1.

This phenomenon is a demonstration of the Central Limit Theorem (CLT) in action. The CLT states that the sum (or equivalently, the average) of a large number of independent, identically distributed random variables tends to follow a normal distribution as the sample size increases – regardless of the original distribution generating said random variables (as long as it has finite mean and variance).

The experiment above provides empirical evidence for this. Even though uniform random variables are not normally distributed, their averages as sample size $n = 100$ already seem to be distributed roughly normal.

While a mathematical proof of the CLT is beyond our scope as it is highly technical, it remains one of the most powerful and useful results in statistics (see [Wikd] for a proof of classical CLT using characteristic functions). It allows us to make probabilistic statements about averages, even when the underlying data distribution is unknown.

Furthermore, the reason for the specific mathematical form of the normal distribution is that it uniquely arises as the limit of sample averages under the Central Limit Theorem – as long as the individual random variables have finite mean and variance. This is why the bell curve shows up so universally across the sciences – and more broadly, in nature.

3.10.3 Birthday problem

Now consider a classic example where the analytical solution is complicated, but the simulation is simple: the **Birthday Paradox**.

We want to compute:

- What is the probability that, in a group of n randomly chosen people, at least two share a birthday – for different group sizes n? For example, the probability of at least two people sharing a birthday in a class of 60 people.

People usually expect this probability to be low, but it is not, it exceeds 50% with just 23 people – a surprising results – so the problem is sometimes called a paradox. See more here: [Wika; BH19].

Let us find the answer to this question using simulation – that is, by generating some random data.

Simulate a group of n people:

```
np.random.seed(999)
n = 60 # group size
# sample a day of birth (from 365) for each person
# all days are assumed to be equally likely (not actually
  ↳  true in the real world)
birthdays = np.random.choice(365, n, replace=True)
birthdays.reshape((6,-1)) # a single sample of birthdays in a
  ↳  group of 60 people
```

```
array([[348, 357, 200, 225, 217,  16, 117, 264, 136, 176],
       [ 69,  66,  11,  11, 149, 228, 250, 306, 318, 327],
       [358, 190, 115, 358,  43,  32, 105, 132,  13, 299],
```

```
     [ 84,  72, 238, 283,  23,  84, 133, 299, 337, 252],
     [216, 313, 177,  43, 257, 234, 206, 259, 364,   9],
     [250, 363,  78,  16, 127, 314,  31, 302,  52, 204]])
```

Check if there is at least one coincidence / clash in birthdays (any duplicates):

```
# check if two or more birthdays coincide -- i.e., we get a
  ↪ hit
def hit(b):
    return 1.0*(np.unique(b).shape[0] < b.shape[0])

hit(birthdays)  # there is a hit! - some birthdays coincide
```

```
1.0
```

Probability (relative frequency) in a group of 60 people of at least two people sharing a birthday via simulation:

```
# simulate 10000 classrooms and measure proportion with at
  ↪ least one coincidence
hits = [hit(np.random.choice(365, n, replace=True)) for i in
  ↪ range(10000)]
prob = np.mean(hits) # probability of at least one birthday
  ↪ coincidence
prob
```

```
np.float64(0.9948)
```

Let us generalize to compute the probability for any group size n:

```
def prob_clash(n):
    hits = [hit(np.random.choice(365, n, replace=True)
               ) for i in range(10000)]
    prob = np.mean(hits)
```

```
    return prob

print("Group size: 22, Prob(coincidence):", prob_clash(22))
print("Group size: 23, Prob(coincidence):", prob_clash(23))


Group size: 22, Prob(coincidence): 0.4724
Group size: 23, Prob(coincidence): 0.51
```

Evaluate for a wide range of group sizes n:

```
np.random.seed(999)
probs = [prob_clash(n) for n in range(5, 80, 5)]
plt.plot(range(5, 80, 5), probs) # group sizes between 5 and
  ↳  80 in 5 int intervals
plt.xlabel("Group size")
plt.ylabel("P(coincidence)")
plt.grid(True, linestyle=':', linewidth=0.5)
plt.show()
```

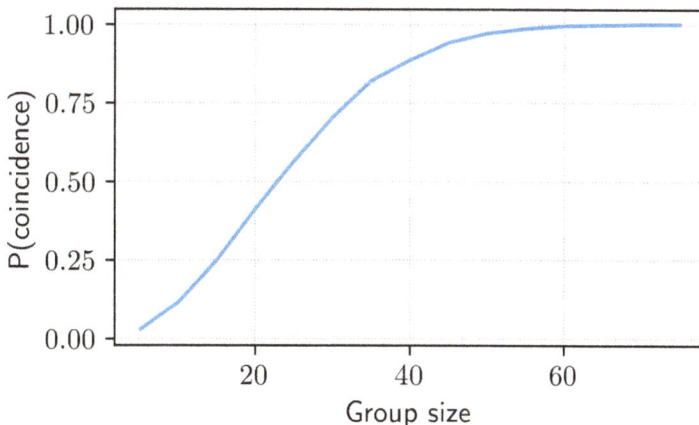

Figure 3.6: Probability of at least one birthday coincidence by group size

This example shows how we can use Monte Carlo simulation – random sampling + aggregation – to estimate complex probabilities without solving equations by hand. Simulation allows us to find the answer to the birthday problem fast without too much effort. Analytical derivation of this result is possible, but is non-trivial and, arguably, more cumbersome. This way simulation sometimes allows us to avoid doing painful math. It demonstrates the power of simulation for intuition and estimation, even when formal math is hard or intractable.

3.11 Discussion

In this chapter, we have observed how probabilities can be used to quantify uncertainty and how computer simulation driven by pseudorandom number generation can be used to solve analytical problems. We will see later how probabilities can be used in machine learning to formulate predictions, reason about prediction quality, and generate / sample new data. The random sampling based on probabilistic predictions turns out to be exactly how language models operate.

3.12 Further learning resources

- Introductory probability: [BH19] and [Ros13].
- Simulation: [Law13] and [Fla+92].
- Measure theory (advanced):
  - Real analysis foundations: [Rud76] and Francis Su's video lectures (https://analysisyawp.blogspot.com/).
  - Measure-theoretic probability: Stanford STAT310A lecture notes (https://web.stanford.edu/class/stats310a/lnotes.pdf) and [Axl20].

3.13 References

[Axl20]   Sheldon Axler. *Measure, Integration & Real Analysis*. Springer Nature, 2020.

[BH19]   Joseph K Blitzstein and Jessica Hwang. *Introduction to Probability*. Chapman and Hall/CRC, 2019.

[Fla+92]   Brian P Flannery et al. "Numerical recipes in C". In: *Press Syndicate of the University of Cambridge, New York* 24.78 (1992), p. 36.

[Law13]   Averill M. Law. *Simulation Modeling and Analysis*. McGraw-Hill, 2013.

[Num]   NumPy. *Random Generator*. URL: https://numpy.org/doc/2.1 /reference/random/generator.html (visited on 05/02/2025).

[Orb]   Peter Orbanz. *Review: Gaussian Distributions*. URL: https://w ww.gatsby.ucl.ac.uk/~porbanz/teaching/UN3106S18/slides_1 8Jan.pdf (visited on 05/06/2025).

[Ros13]   Sheldon Ross. *A First Course in Probability*. Pearson Higher Ed, 2013.

[Rud76]   Walter Rudin. *Principles of Mathematical Analysis*. McGraw-Hill, Inc, 1976.

[Wika]   Wikipedia. *Birthday problem*. URL: https://en.wikipedia.org /wiki/Birthday_problem (visited on 05/02/2025).

[Wikb]   Wikipedia. *Cantor distribution*. URL: https://en.wikipedia.org /wiki/Cantor_distribution (visited on 05/30/2025).

[Wikc]   Wikipedia. *Cantor's diagonal argument*. URL: https://en.wiki pedia.org/wiki/Cantor%27s_diagonal_argument (visited on 05/30/2025).

[Wikd]   Wikipedia. *Central limit theorem*. URL: https://en.wikipedia.o rg/wiki/Central_limit_theorem (visited on 05/02/2025).

[Wike]   Wikipedia. *Conditional probability distribution*. URL: https://e n.wikipedia.org/wiki/Conditional_probability_distribution (visited on 05/02/2025).

[Wikf]   Wikipedia. *Convolution of probability distributions*. URL: http s://en.wikipedia.org/wiki/Convolution_of_probability_distri butions (visited on 05/30/2025).

[Wikg]   Wikipedia. *Law of large numbers*. URL: https://en.wikipedia.o rg/wiki/Law_of_large_numbers (visited on 05/02/2025).

[Wikh]   Wikipedia. *Linear congruential generator*. URL: https://en.wi kipedia.org/wiki/Linear_congruential_generator (visited on 05/02/2025).

[Wiki]   Wikipedia. *List of probability distributions*. URL: https://en.w ikipedia.org/wiki/List_of_probability_distributions (visited on 05/30/2025).

[Wikj]   Wikipedia. *Mersenne Twister*. URL: https://en.wikipedia.org /wiki/Mersenne_Twister (visited on 05/02/2025).

[Wikk]   Wikipedia. *Moment*. URL: https://en.wikipedia.org/wiki/Mo ment_(mathematics) (visited on 05/02/2025).

[Wikl]   Wikipedia. *Moment-generating function*. URL: https://en.w ikipedia.org/wiki/Moment-generating_function (visited on 05/30/2025).

[Wikm]   Wikipedia. *Non-uniform random variate generation*. URL: htt ps://en.wikipedia.org/wiki/Non-uniform_random_variate_g eneration (visited on 05/30/2025).

[Wikn]   Wikipedia. *Probability distribution*. URL: https://en.wikipedia .org/wiki/Probability_distribution (visited on 05/30/2025).

4 Regression models

In this chapter, we review regression analysis – a class of algorithms used to predict an output variable from one or more input variables. We will cover two specific types: (1) linear regression for predicting continuous outcomes, and (2) multinomial logistic regression for discrete, multiple-choice-style prediction. Understanding regression analysis lays the groundwork for grasping how neural networks and large language models operate, as these complex systems are built from components – called neurons – that echo the structure and logic of simple regression models.

4.1 Linear regression predictor

Suppose we have data on the price and weight of *diamonds*. Our goal is to predict a diamond's price based on its weight. This could be a useful benchmark and sanity check for pricing decisions.

We can visualize the data as a 2D scatter plot (that is, a 2D geometric space), with the weight in carats on the x-axis (1 carat = 0.2 grams), and the price in US dollars on the y-axis (see Figure 4.1). Each diamond in the data is a point on this scatter plot, with its weight and price being its coordinates. A simple prediction approach involves drawing a straight line through the point cloud. To estimate the price for a given carat weight, we locate the corresponding value on the x-axis, move upward until we intersect the regression line, and then read the predicted price by moving horizontally to the y-axis. This line is a basic prediction *model* – a simplified mathematical representation of the relationship between variables – in this case, the diamond price and carat weight.

The fit of this prediction line, also known as the regression line, to the data can be assessed by measuring the vertical distance between each actual data point and the corresponding price value predicted by the line. For instance, if a point representing an actual diamond lies below the line, it indicates the model has overestimated the price for that diamond. Conversely, if the point is above the line, the actual price exceeds the model's prediction. Because this model only considers weight and ignores other influential factors such as clarity and color, such deviations are expected. These vertical differences between true and predicted values are known as *residuals* or *errors*.

Figure 4.1 illustrates our current setup:

- The blue line represents the *regression line*, which provides predicted prices $\hat{y}$ (pronounced *y hat*) for input carat weight.
- The vertical dashed red lines represent *residuals*: the differences $y_i - \hat{y}_i$ between actual and predicted values for specific stones, indexed by i, in the data.

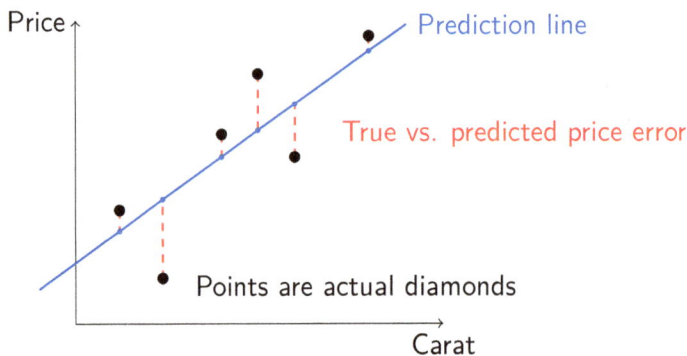

Figure 4.1: Actual vs. predicted diamond prices – illustration

We could draw infinitely many lines through this cloud of points. Among many possible lines, we seek the one that best fits the data. To compare lines, we need a numerical criterion to quantify prediction quality. The difference between true and predicted prices for each data point is a good candidate.

Let y_i be the true price and $\hat{y}_i$ the predicted price for diamond i. The prediction error ϵ_i is

$$\text{error}_i = \epsilon_i = y_i - \hat{y}_i.$$

(Greek letter ϵ is pronounced as *epsilon*.) Errors can cancel out if summed directly – because their signs can be either negative or positive. What we really should focus on are the error magnitudes. Taking the absolute error value is one option, but squaring the errors is more mathematically convenient and leads to some useful analytics properties. Averaging squared errors across n observed stones gives us the *mean squared error (MSE)*:

$$\text{MSE} = \frac{1}{n} \sum_{i=1}^{n} (y_i - \hat{y}_i)^2.$$

Due to squaring, whether we compute $y_i - \hat{y}_i$ or $\hat{y}_i - y_i$ makes no difference to the MSE value.

MSE serves as a useful metric for comparing prediction lines. We thus agree that the best-fitting line is the one that minimizes the MSE.

---

i Weighted errors

In some applications, we may care about errors of different signs to a different degree – and we could put weights on such different types of errors – but we do not worry about this nuance here.

Next, we describe this setup mathematically, paving the way to identifying the best prediction line.

4.1.1 Mathematical representation of a line

Let β_0 and β_1 (denoted by the Greek letter *beta*) be two scalar numbers, called parameters. The carat-to-price prediction line is given by equation $\hat{y} = \beta_0 + \beta_1 \text{carat}$. For a diamond i, the prediction is $\hat{y}_i = \beta_0 + \beta_1 \text{carat}_i$. Note that β_0 and β_1 fully determine the prediction line.

Let us try to better understand the meaning of β_0 and β_1 parameters:

- **Intercept term** β_0. When carat $= 0$ (diamond of zero weight), $\hat{y}_i = \beta_0 + \beta_1 \text{carat}_i$ becomes $\hat{y}_i = \beta_0$. So β_0 indicates the price point where prediction line intersects the y-axis, as this is where carat $= 0$ – see the Figure 4.1 scatter plot to make sure you see why this is. While we would expect the price of a zero-weight diamond to be zero, a potentially non-zero value could result from extrapolation, as no zero-carat diamonds exist in the data.

- **Line slope** β_1. Increasing the carat by 1 increases the predicted price by β_1 units: $\beta_0 + \beta_1(\text{carat} + 1) = \beta_0 + \beta_1\text{carat} + \beta_1$. Thus, β_1 is the sensitivity of diamond price to changes in weight – "by how much does output change with a unit increase in input." It is also exactly the definition of a slope of a line $\Delta\text{price}/\Delta\text{carat}$, where Δprice, pronounced *delta price* (denoted by the Greek capital letter *delta*), stands for the change in the price corresponding to the change in the carat weight Δcarat. (Delta symbol Δ is commonly used to indicate the change in the variable that follows it; it can also be used on its own to indicate a quantity by which some variable changes (e.g., $x + \Delta$).)

Given this formulation, the prediction error for each diamond becomes:

$$\text{error}_i = \epsilon_i = \underbrace{y_i}_{\text{true price}} - \underbrace{\hat{y}_i}_{\text{predicted price}} = y_i - \beta_0 - \beta_1\text{carat}_i.$$

Thus, MSE can be written as:

$$\text{MSE} = \frac{1}{n}\sum_{i=1}^{n}(y_i - \beta_0 - \beta_1\text{carat}_i)^2.$$

This equation captures MSE's dependence on y_i and carat_i, which are fixed and given in the data, and β_0, β_1 that define the line. We can now search for values of β_0 and β_1 that minimize the MSE.

This logic generalizes to multiple input variables. For example, adding the diamond clarity as another input gives $\hat{y} = \beta_0 + \beta_1\text{carat} + \beta_2\text{clarity}$, where β_2 is a new parameter capturing the change in price with a unit increase in clarity. The corresponding MSE becomes $\text{MSE} = \frac{1}{n}\sum_{i=1}^{n}(y_i - \beta_0 - \beta_1\text{carat}_i - \beta_2\text{clarity}_i)^2$.

Geometrically, prediction error would now be not a vertical distance between a point and a line in 2D space, but instead a vertical distance between a point and a 2D plane in 3D space – with 2D plane's orientation determined by β_0, β_1, β_2 parameters. Our graphical visualization above would become 3D as we would introduce a third axis for the clarity variable, going orthogonally to the price-carat plane. Figure 4.2 shows an outline of such a 3D space.

Additional input variables expand this to higher dimensions, which becomes harder to visualize but is easily handled algebraically.

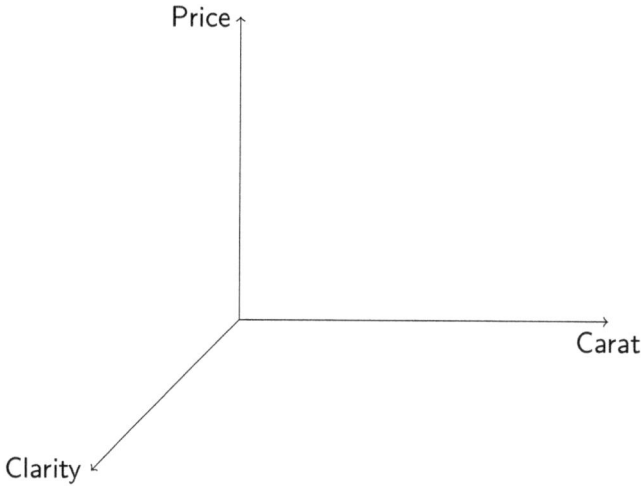

Figure 4.2: Example of a 3D space of diamonds

4.1.2 Matrix notation for linear regression

Matrix notation simplifies our equations and is well-suited for implementation. First, consider our earlier prediction equation $\beta_0 + \beta_1 \text{carat}_i$. We will now put all the parameters into a single vector

$$\beta = \begin{bmatrix} \beta_0 \\ \beta_1 \end{bmatrix}.$$

Here β is shaped 2×1.

We can re-write the above equation, explicitly incorporating a scalar 1 (intercept), $\beta_0 + \beta_1 \text{carat}_i = \beta_0 \cdot 1 + \beta_1 \text{carat}_i$ without changing the results.

We can then write an input vector

$$x = \begin{bmatrix} 1 \\ \text{carat}_i \end{bmatrix}$$

and write the prediction using vector product as $\beta_0 \cdot 1 + \beta_1 \text{carat}_i = \beta^T x$.

The full input matrix X for all n diamonds, where each row is a different diamond, is $n \times 2$:

$$X = \begin{bmatrix} 1 & \text{carat}_1 \\ 1 & \text{carat}_2 \\ \vdots & \vdots \\ 1 & \text{carat}_n \end{bmatrix}.$$

The true price $n \times 1$ output vector y is

$$y = \begin{bmatrix} y_1 \\ y_2 \\ \vdots \\ y_n \end{bmatrix}$$

and the $n \times 1$ predicted price vector $\hat{y}$ is

$$\hat{y} = X\beta = \begin{bmatrix} 1 & \text{carat}_1 \\ 1 & \text{carat}_2 \\ \vdots & \vdots \\ 1 & \text{carat}_n \end{bmatrix}_{n \times 2} \cdot \begin{bmatrix} \beta_0 \\ \beta_1 \end{bmatrix}_{2 \times 1} = \begin{bmatrix} \beta_0 + \beta_1 \text{carat}_1 \\ \beta_0 + \beta_1 \text{carat}_2 \\ \vdots \\ \beta_0 + \beta_1 \text{carat}_n \end{bmatrix}_{n \times 1}.$$

Vector of residuals / errors for true vs. predicted prices is just

$$\epsilon = y - \hat{y} = \underbrace{y}_{n \times 1} - \underbrace{X\beta}_{n \times 1} = \begin{bmatrix} y_1 - \beta_0 - \beta_1 \text{carat}_1 \\ y_2 - \beta_0 - \beta_1 \text{carat}_2 \\ \vdots \\ y_n - \beta_0 - \beta_1 \text{carat}_n \end{bmatrix}_{n \times 1}.$$

The MSE formula in matrix notation is

$$\begin{aligned} \text{MSE} &= \frac{1}{n}(y - \hat{y})^T(y - \hat{y}) \\ &= \frac{1}{n}(y - X\beta)^T(y - X\beta) \\ &= \text{mean}((y - X\beta)^2) \\ &= \mathbf{1}^T(y - X\beta)^2/n, \end{aligned}$$

where $\underbrace{(y - X\beta)^T}_{1 \times n} \underbrace{(y - X\beta)}_{n \times 1}$ operation squares each element in residual vector $(y - X\beta)$ and adds them up; $\mathbf{1}$ is $n \times 1$ vector of ones – here, dot product by it performs summation; $(X\beta - y)^2$ is a vector of elementwise-squared residuals.

To highlight the dependence on model parameters, we can write the mean squared error explicitly as a function of β:

$$\text{MSE}(\beta) = \text{MSE}(\beta_0, \beta_1) = \frac{1}{n}(y - X\beta)^T(y - X\beta).$$

Here, the true output data y and input data X are fixed and known. The only quantities we vary to reduce MSE are the components of β, which define the prediction line.

A key advantage of this matrix formulation is its generality – it remains valid regardless of the number of observations or the number of input variables – making it well-suited for implementation in code.

4.2 Finding the optimal linear predictor using derivatives

Let us summarize where we are.

We want to predict a vector y of n values (e.g., prices). We assume a linear relationship $y \approx \hat{y} = X\beta$, where X is an $n \times k$ matrix of input variables (including the intercept – a column vector of n ones), and β is a vector of k parameters that control how input variables form the prediction vector $\hat{y}$ of n values. The prediction quality is measured by the mean squared error (MSE):

$$\text{MSE}(\beta) = \frac{1}{n}(y - X\beta)^T(y - X\beta).$$

Our goal is to find a parameter vector β^* that minimizes this MSE:

$$\beta^* = \arg\min_{\beta} \text{MSE}(\beta),$$

where $\arg\min$ stands for *argument of the minimum.*

Here are two possible strategies:

- **Random search**. One simple approach is to use a random number generator to sample many candidate values for β, compute the MSE for each, and retain the one with the lowest error. With enough samples, this can yield a reasonably good solution. However, pinpointing the exact optimal values would require more targeted strategies – such as focusing the sampling around promising regions of the parameter space. Moreover, as the number of parameters increases, the search space grows exponentially, making it increasingly inefficient to cover all possible combinations through random sampling alone.
- **Derivatives**. Alternatively, we can rely on calculus – specifically, derivatives – to guide us towards the optimal parameter vector β^*.

4.2.1 A refresher on derivatives

We start with a basic example function: $f(x) = x^2$. The **derivative** of a function $f(x)$ at a point x, written as $\frac{df}{dx}$ (alternatively, $f'(x)$), tells us *how the function changes when the input changes slightly $x \to x + \Delta$*:

$$\frac{df}{dx} = \lim_{\Delta \to 0} \frac{f(x + \Delta) - f(x)}{\Delta},$$

where $\lim_{\Delta \to 0}$ means the limit of the expression as Δ (change to x) goes to zero.

For $f(x) = x^2$, we get:

$$\begin{aligned}
\frac{df}{dx} &= \lim_{\Delta \to 0} \frac{f(x + \Delta) - f(x)}{\Delta} \\
&= \lim_{\Delta \to 0} \frac{(x + \Delta)^2 - x^2}{\Delta} \\
&= \lim_{\Delta \to 0} \frac{x^2 + 2x\Delta + \Delta^2 - x^2}{\Delta} \\
&= \lim_{\Delta \to 0} (2x + \Delta) = 2x.
\end{aligned}$$

Note:

- The first equality is the definition of derivative.
- The second equality is us plugging in $f(x) = x^2$.
- The third and fourth equalities is opening up the brackets and some algebraic elimination.

- Taking the limit $\lim_{\Delta \to 0}$ allows us to eliminate Δ in the final expression.

Thus, for $f(x) = x^2$, the derivative is $\frac{df}{dx} = 2x$, as you may remember from the high school calculus. In words: starting at some point x, increasing x by some small amount Δ changes $f(x)$ by approximately $2x \cdot \Delta$.

To emphasize, the derivative itself is often a function of the parameter – we could optionally write $\frac{df}{dx} = \frac{df}{dx}(x) = 2x$. That is, the derivative value is different for different x – *how $f(x)$ changes when x is tweaked, in general, depends on the starting point x.*

The function plot for $f(x) = x^2$ is provided in Figure 4.3. Geometrically, the derivative of $f(x)$ at point x is the slope of the tangent line to the function curve at that point. For example, with $f(x) = x^2$ and $\frac{df}{dx}(x) = 2x$, at $x = 1$, the derivative is positive $\frac{df}{dx}(1) = 2$, indicating a positive slope – and that increasing x will increase the function when starting at $x = 1$, while reducing x should reduce $f(x)$. At $x = -1$, the derivative is negative $\frac{df}{dx}(-1) = -2$, indicating a negative slope and that increasing x from $x = -1$ will reduce the function.

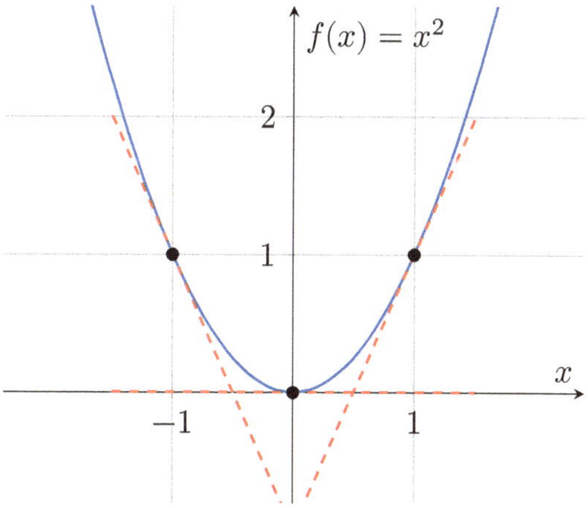

Figure 4.3: A quadratic function

At $x = 0$, the derivative is zero $\frac{df}{dx}(0) = 0$, indicating a zero slope (horizontal tangent line). When the derivative is zero at a point x_0, we expect the function not to change much when tweaking x_0. Such points are called *stationary points*. A stationary point often suggests the presence of a *local extremum* – that is, a *local* minimum or maximum value relative to nearby points. In the case of $f(x) = x^2$, $x = 0$ indeed yields the minimum value of the function; in fact, it is a *global* minimum point – no other x value in the whole domain of $f(x)$ can yield a lower $f(x)$ value. (The *domain* of a function is the complete set of input values for which the function is defined and produces a valid output.)

If the function has more than one variable, say $f(x, y) = x^2 + y^2$, we can compute **partial derivatives**:

$$\frac{\partial f}{\partial x} = 2x, \quad \frac{\partial f}{\partial y} = 2y.$$

---

i Total vs. partial derivative notation

By convention, for single-variable functions, the derivative notation is $\frac{df}{dx}$, known as the *total* derivative; for functions of multiple variables, the derivative notation involves a slightly different expression $\frac{\partial f}{\partial x}$, known as the *partial* derivative.

When we compute such a partial derivative of a function on a specific input variable, we only change that input variable – holding the other input variables constant. For example:

$$\begin{aligned}
\frac{\partial f}{\partial x} &= \lim_{\Delta \to 0} \frac{f(x + \Delta, y) - f(x, y)}{\Delta} \\
&= \lim_{\Delta \to 0} \frac{(x + \Delta)^2 + y^2 - x^2 - y^2}{\Delta} \\
&= \lim_{\Delta \to 0} \frac{(x + \Delta)^2 - x^2}{\Delta} \\
&= (\text{some algebra...}) = 2x.
\end{aligned}$$

This tells us that as x increases by small Δ, the function increases approximately by $2x \cdot \Delta$ – and y variable does not affect the magnitude of this change. We could similarly compute $\frac{\partial f}{\partial y} = \lim_{\Delta \to 0} \frac{f(x, y+\Delta) - f(x, y)}{\Delta} = 2y$.

The **gradient** of a function f is denoted by ∇f – the symbol *nabla* ∇ concatenated with the function's name – and is simply the array of all function's partial derivatives, shaped identically to the array of input variables that the derivatives are computed with respect to.[1] For our example function $f(x, y) = x^2 + y^2$, input array is 2×1:

$$\begin{bmatrix} x \\ y \end{bmatrix},$$

and the gradient is thus

$$\nabla f(x, y) = \begin{bmatrix} \frac{\partial f}{\partial x} \\ \frac{\partial f}{\partial y} \end{bmatrix} = \begin{bmatrix} 2x \\ 2y \end{bmatrix}.$$

As with specific partial derivatives, gradient $\nabla f(x, y)$ is a function of specific parameter values x and y and gives us an idea of how $f(x, y)$ changes if we increase x or y a little bit from some given starting values for x and y.

Interpretation. Starting with some specific numbers x and y, the gradient (partial derivatives) tells us the direction in the input variable space in which the function increases / decreases. A *positive* partial derivative value at a point x_0 means the function will *increase* if we increase the input value a bit. A *negative* partial derivative value at a point x_0 means the function will *decrease* if we increase the input value a bit. So, to minimize a function, we tweak input variables x and y to move in the opposite direction of the gradient.

For instance, if $\frac{\partial f}{\partial x}(x_0, y_0)$ is large and positive at our initial set of points x_0 and y_0, this means that increasing x_0 will further increase $f(x, y)$, while decreasing x_0 should instead make the function smaller – which is what we would want to do if the function is MSE!

And if the partial derivative $\frac{\partial f}{\partial x}(x_0, y_0)$ is equal to zero, that means shifting the parameter x_0 just a little bit won't change the function much. For a general function, when the gradient is zero (i.e., all partial derivatives are zero), we are at a *stationary point* – shifting any of the parameters just a little bit won't change the function much. This signals the function might be at a local minimum or maximum. Sometimes, we can analytically find such extreme function points, by solving equations, after setting all partial derivatives to zero. We could then check if any of the stationary points indeed correspond to the function's extreme point that we are seeking (e.g., a global minimum).

---

[1]Do not confuse the *nabla* ∇ gradient symbol with the *delta* Δ symbol for change.

4.2.2 Gradient descent

This insight leads to the *gradient descent algorithm*: an iterative method for finding a local minimum of a function by subtracting from the current parameter estimate the derivative of the function evaluated at that estimate. For a function of one variable, the update rule is:

$$x_{t+1} = x_t - \gamma \cdot \frac{df}{dx}(x_t).$$

Here γ (pronounced *gamma*) is a **learning rate**, a small positive number controlling the step size. Subscript t indicates time: x_t is parameter estimate at time t; x_{t+1} is the new parameter estimate at time $t + 1$.

Explanation: If the derivative $\frac{df}{dx}(x_t)$ is positive at the current estimate x_t, this tells us that decreasing x_t should decrease the function. Accordingly, this rule subtracts this positive derivative from x_t, reducing x_t, so value for $f(x_{t+1})$ should be smaller than $f(x_t)$. Alternatively, if the derivative is negative at the current guess of x_t then we need to increase x_t in order to shrink the function – then the update rule subtracts the negative number for the derivative from x_t – and so actually increases x_t, shrinking the function value as intended.

For example, with $f(x) = x^2$, $\frac{df}{dx}(x) = 2x$, and $\gamma = 0.001$, if $x_0 = 1$, then:

$$x_1 = 1 - 0.001 \cdot 2 = 0.998.$$

Each step moves closer to the function's minimum at $x = 0$.

Alternatively, if $x_0 = -1$, $\frac{df}{dx}(-1) = -2$, and $x_1 = -1 - 0.001 \cdot (-2) = -1 + 0.002 = -0.998$. So we are also closer to the function's minimum at $x = 0$.

If we are at $x_t = 0$, derivative at that point is 0, so the rule would not change the estimate further.

Small positive γ is critical to ensure smooth progress. If we had $\gamma = 1$ and made the update from $x_0 = 1$, we would get $x_1 = x_0 - 2 = 1 - 2 = -1$. So instead of slowly approaching $x = 0$, we would overshoot in the update, oscillating between $x = 1$ and $x = -1$. The gradient descent would *not converge* to the solution in this case.

In higher dimensions – in case of multivariable functions – we update all parameters simultaneously:

$$\beta_{t+1} = \beta_t - \gamma \cdot \nabla_\beta \text{MSE}(\beta_t).$$

In $\nabla_\beta \text{MSE}(\beta_t)$, β subscript in ∇_β gradient operator highlights that the gradient of the function MSE is taken with respect to the parameter vector β – and not with respect to some other inputs to MSE (such as input data X and y, for example – as we could ask how sensitive MSE is with respect to changes in those inputs too). Sometimes, the subscript in the gradient operator can be omitted when the context clearly indicates which variable is being differentiated. For example, it is common to see $\nabla \text{MSE}(\beta)$ instead of $\nabla_\beta \text{MSE}(\beta)$ because it is generally understood that X and y are treated as fixed data and that β is the parameter vector we are optimizing – as highlighted by the bracket notation for the argument (β).

For example, for $f(x, y) = x^2 + y^2$, we get the gradient descent update:

$$\begin{bmatrix} x_{t+1} \\ y_{t+1} \end{bmatrix} = \begin{bmatrix} x_t \\ y_t \end{bmatrix} - \gamma \cdot \nabla f(x_t, y_t) = \begin{bmatrix} x_t \\ y_t \end{bmatrix} - \gamma \cdot \begin{bmatrix} \frac{\partial f}{\partial x}(x_t, y_t) \\ \frac{\partial f}{\partial y}(x_t, y_t) \end{bmatrix} = \begin{bmatrix} x_t \\ y_t \end{bmatrix} - \gamma \cdot \begin{bmatrix} 2x_t \\ 2y_t \end{bmatrix}.$$

4.2.3 MSE gradient

Let us now apply this to our linear regression problem. MSE is a function of the parameters β and so can be differentiated with respect to those parameters to get the gradient vector $\nabla_\beta \text{MSE}(\beta)$, which itself is a function of β. Recall:

$$\text{MSE}(\beta) = \frac{1}{n}(X\beta - y)^T(X\beta - y).$$

Opening the brackets, using rules of transpose of matrix dot product (to use the fancy language, we *expand the quadratic form*):

$$\text{MSE}(\beta) = \frac{1}{n}(\beta^T X^T X \beta - \beta^T X^T y - y^T X \beta + y^T y).$$

As $\beta^T X^T y = y^T X \beta$ (both expressions evaluate to a scalar value):

$$\text{MSE}(\beta) = \frac{1}{n}(\beta^T X^T X \beta - 2\beta^T X^T y + y^T y).$$

Notice here that the MSE is a *quadratic function* of the parameters β (resembling Figure 4.3, but generalized to multiple dimensions).

Taking the derivative with respect to vector β:

$$\nabla_{\beta}\text{MSE}(\beta) = \frac{1}{n}(2X^TX\beta - 2X^Ty)$$
$$= \frac{2}{n}X^T(X\beta - y).$$

This gradient vector has the same shape as β parameter vector $(k \times 1)$ – and tells us how MSE changes with change in each number in the parameter vector β.

4.2.4 Two ways to use the MSE gradient

A. Analytical solution

Knowing the gradient expression, we can set the derivatives to zero and solve for β. The fact that MSE is a quadratic function of β ensures that setting its gradient to zero will identify its global minimum (no other β can yield lower MSE); so we do not have to worry about other kinds of stationary points (e.g., maximum). We get:

$$\nabla_{\beta}\text{MSE}(\beta) = \frac{2}{n}X^T(X\beta - y) = 0$$
$$X^TX\beta = X^Ty,$$

which gives us an analytical solution for optimal parameters, called **ordinary least squares (OLS)**:

$$\beta^* = (X^TX)^{-1}X^Ty.$$

This unique closed-form solution exists as long as X^TX is invertible – the determinant of X^TX is non-zero. This β^* solution identifies the unique minimum of the MSE.

Multicollinearity. One common cause of a non-invertible matrix X^TX is the presence of a duplicated column in X – a case of *perfect multicollinearity*. In this case, there is not a unique MSE-minimizing β vector – multiple parameter combinations can produce the same model predictions and equally low MSE. We say the parameters are *not identifiable*. For example, consider estimating three coefficients in a regression equation:

$$\beta_0 + \beta_1 x_a + \beta_2 x_b,$$

where, as it turns out, $x_a = x_b$. Then:

$$\beta_0 + \beta_1 x_a + \beta_2 x_b = \beta_0 + \beta_1 x_a + \beta_2 x_a = \beta_0 + (\beta_1 + \beta_2)x_a.$$

Here, infinitely many combinations of β_1 and β_2 can yield the same slope for x_a and thus the same prediction, so there is no unique optimal solution in terms of β. Even when the variables are not exactly collinear but are highly correlated, such imperfect multicollinearity can make $X^T X$ *ill-conditioned*, resulting in numerically unstable coefficient estimates.

One simple solution is to keep only one among the duplicated / very highly correlated variables in the data. However, removing a variable that is only highly correlated (not perfectly) discards some information, increasing the MSE on the train data. Another way $X^T X$ becomes non-invertible is if the number of rows (observations) in X is less than the number of its columns (variables). Later in this chapter, we will see a distinct approach, called *regularization*, which addresses non-identifiability problems by modifying the objective function – and also serves as a universal remedy for the non-invertibility of $X^T X$.

B. Gradient descent

We can also use the gradient $\nabla_\beta \mathrm{MSE}(\beta_t)$ to search for β^* that minimizes MSE iteratively:

$$\beta_{t+1} = \beta_t - \gamma \nabla_\beta \mathrm{MSE}(\beta_t) = \beta_t - \gamma \cdot \frac{2}{n} X^T (X\beta_t - y).$$

This method is especially useful when:

- The analytical solution is computationally expensive (e.g., for very large datasets).
- The model is non-linear and there is no closed-form solution.

Gradient descent gradually adjusts β to minimize MSE. If the learning rate γ is chosen appropriately (typically, small), gradient descent applied to MSE is guaranteed to converge to a global minimum (since this MSE is a quadratic function of β). If $X^T X$ is invertible, this global minimum is unique. Otherwise, there are many minimizers, and gradient descent will land on one of them.

4.3 Sidebar: General vs. convex optimization problems

Consider an example of a general single-variable non-linear function in Figure 4.4 that one might be interested in optimizing.

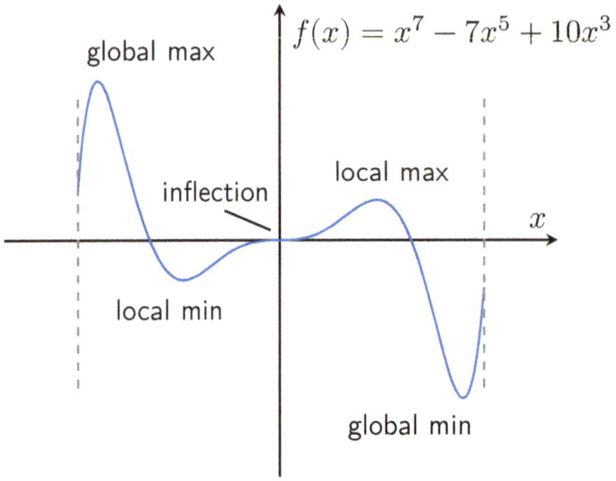

Figure 4.4: A non-linear function defined on a bounded domain $x \in [-2.2; 2.2]$, with multiple stationary points

For such general functions, setting their gradients to zero can identify different kinds of stationary points:

- **Global minimum (point)**: A point where the function attains its lowest possible value across the entire domain (domain is the set of all input values for which the function is defined). Also called a global *minimizer*. Formally, a point x_{gm} is a global minimum if $f(x_{gm}) \leq f(x)$ for all x in the domain. No other input yields a smaller function value. A global minimum may not be unique – multiple inputs can yield the same lowest value. A global minimum may not exist at all – e.g., $f(x) = x$ is unbounded below and has no global minimum (the point has to be a real number, $-\infty$ is not a valid point as it is not a part of the real number line).
- **Local minimum (point)**: A point where the function achieves the *lowest value within a local neighborhood*. Formally, x_{lm} is a local minimum if there exists $\Delta > 0$ such that $f(x_{lm}) \leq f(x)$ for all $x \in (x - \Delta; x + \Delta)$. That is, no nearby point produces a lower output. Local minima may not be unique.

- **Note:** Every global minimum is also a local minimum, but not every local minimum is a global one.

- **Global / local maximum (points):** Analogously, points where the function attains its highest value globally / locally.

  - **Important!** Minimizing any function $f(x)$ is *equivalent* to maximizing its negative $-f(x)$, and vice versa. Formally, $\arg\min_x f(x) = \arg\max_x -f(x)$; that is, the set of global minima of $f(x)$ equals the set of global maxima of $-f(x)$. In practice, many optimization algorithms are designed for the default of minimization (e.g., gradient *descent*), so if you want to maximize some function, you often minimize its negative.

- **Saddle point (or point of inflection):** A stationary point that is neither a local minimum nor a local maximum. Although the gradient is zero, the function does not achieve an extremum at that point. Instead, the function may increase in some directions and decrease in others – exhibiting a "flat spot" with mixed curvature behavior.

Additionally, when the set of allowed parameter values is constrained, the function may attain its minimum or maximum not only at stationary points but also at the **boundary points** of the domain. These boundary points must therefore be examined as additional candidates for extrema.

> **i** Point or output?
>
> The term *global minimum* is commonly used to refer to the global minimum *point*; however, occasionally, the term may be used to refer to the global minimum *value* that the function yields at the global minimum point. Further, the term can also be used to refer to both the point and the value assumed there. The same ambiguity occurs with other stationary points. The intended meaning should usually be clear from the context.

For general functions, gradient descent may converge to a *local minimum*, which is not necessarily a *global minimum*. As a result, there is no guarantee of finding the optimal solution – that is, there could be some better parameter value that the algorithm just failed to find. In these cases, the outcome of optimization can depend heavily on the initial starting point, as different starting values may lead the algorithm to different local minima.

There is, however, a special class of **convex** functions, which all share the property that *any local minimum of a convex function is also a global minimum* – if such a minimum exists. As a result, for a convex function that has a global minimum, any local minimum found by gradient descent is also a global minimum.

Convex functions with respect to the argument x include, for example, affine $f(x) = ax + b$ (a, b are constants), quadratic $f(x) = x^2$, and exponential $f(x) = \exp(x)$ functions. A nonnegative-weighted sum of convex functions of x remains a convex function of x. Convexity is preserved under some other operations too. Linear regression MSE function, as a multivariate quadratic function of vector β, is convex.

Formally, a function is convex if the line segment between any two distinct points on its graph lies above or on the graph between those two points; equivalently, $f(\cdot)$ is convex if and only if, for all valid input points x_1, x_2, $x_1 \neq x_2$, and all $0 < w < 1$, $f(wx_1 + (1-w)x_2) \leq wf(x_1) + (1-w)f(x_2)$.

A *convex optimization problem* is the problem of finding a global minimum of a convex function over a nicely shaped space of possible parameter values, called a feasible *convex set* or *convex region*. A parameter region is called convex if, whenever you pick any two points inside it, the straight line between those two points is also completely inside the region. Convex optimization problems can accommodate useful constraints restricting the allowed parameter values – as long as the constraints preserve the convexity of the feasible region. For example, we could set $x \geq 0$.

Minimum uniqueness. A minimum point of a convex function may not be unique. A convex function that has a unique minimizer is called *strictly convex*. A strictly convex function is defined by replacing $\leq$ with a strict inequality $<$ in the convex function definition above: $f(wx_1 + (1-w)x_2) < wf(x_1) + (1-w)f(x_2)$. It follows that a sum of a convex and a strictly convex functions (defined on the same convex domain) is strictly convex – just add the convex and strictly convex inequalities together and notice that strict inequality $<$ is preserved. This proves to be useful if we want to ensure solution uniqueness for an optimization problem.

Minimum existence. A minimum point of a convex function may not exist.

For example, a convex function $f(x) = x$ is unbounded below (with x unconstrained over the real number line $x \in (-\infty; \infty)$), and a minimum point does not exist. (If x were constrained, e.g., $x \geq 0$, the minimum of $f(x) = x$ would be achieved at $x = 0$ boundary.)

For any convex function $f(x)$, its negative $-f(x)$ is a *concave* function, and vice versa. Maximization of a concave function can thus be trivially converted into minimization of a convex function – by minimizing the negative of the concave function. So, the term *convex optimization* also covers maximization of concave functions.

The advantage of convex optimization problems is that they can usually be solved very efficiently using gradient-descent-like methods – and there are theoretical guarantees that the optimal parameter values found are indeed at a global minimum (if such a minimum exists) – subject to constraints and up to some small error tolerance. Python library **cvxpy** offers a great toolkit for convex optimization – we will see examples later. A detailed discussion of convexity is beyond the scope of this book – see [Wikc; BV04] for more details.

The following example provides a visual comparison of the convex quadratic (in terms of β) linear regression mean squared error (MSE) function and the non-convex Himmelblau's function [Wike], of two variables each, in 3D.

```
import numpy as np
from matplotlib import pyplot as plt


# OPTIONAL: settings to make figures look nicer
# see documentation:
  ↪ https://matplotlib.org/stable/api/rcsetup_api.html
import matplotlib as mpl
from cycler import cycler # color cycling

colors = mpl.color_sequences["petroff10"] #
  ↪ https://arxiv.org/abs/2107.02270
plt.rcParams.update({
    'figure.dpi' : 300,
    'savefig.dpi' : 300,
    'text.usetex' : True, # requires latex installed
```

```python
    'font.size': 11,
    'axes.labelsize': 11,
    'legend.fontsize': 11,
    'axes.prop_cycle' : cycler('color', colors),
    'figure.autolayout': True,
    'savefig.bbox': 'tight',
})


from mpl_toolkits.mplot3d import Axes3D
from matplotlib.colors import LogNorm

# define the convex quadratic function (MSE):
# f(b0,b1) = 0.5(2 - b0 - b1*0.5)^2 + 0.5(-1 - b0 -
  ↪ b1*(-1.5))^2
def convex_quadratic(b0, b1):
    return 0.5*(2 - b0 - b1*0.5)**2 + 0.5*(-1 - b0 -
      ↪ b1*(-1.5))**2

# define the Himmelblau's function:
# f(x, y) = (x^2 + y - 11)^2 + (x + y^2 - 7)^2
def himmelblau(x, y):
    return (x**2 + y - 11)**2 + (x + y**2 - 7)**2

# create a meshgrid
x = np.linspace(-6, 6, 200)
y = np.linspace(-6, 6, 200)
X, Y = np.meshgrid(x, y)

# compute function values
Z1 = convex_quadratic(X, Y)
Z2 = himmelblau(X, Y)


# create the plots
fig, axs = plt.subplots(2, 1, figsize=(5, 10),
  ↪ subplot_kw=dict(projection='3d'))
```

```python
# quadratic plot
surf1 = axs[0].plot_surface(X, Y, Z1, alpha=0.8, cmap='jet',
 ↪ norm=LogNorm())
axs[0].set_title("(a) Linear regression mean squared error
 ↪ (MSE, quadratic and convex):\n"
  r"$MSE(\beta_0,\beta_1) = 0.5(2 - \beta_0 - \beta_10.5)^2 +
  ↪ 0.5(-1 - \beta_0 - \beta_1(-1.5))^2$")
axs[0].set_xlabel(r"$\beta_0$")
axs[0].set_ylabel(r"$\beta_1$")
axs[0].set_zlabel(r"$MSE(\beta_0,\beta_1)$")
fig.colorbar(surf1, ax=axs[0], shrink=0.5, aspect=10,
 ↪ pad=0.15)
axs[0].view_init(elev=50, azim=-30)

# Himmelblau's function plot
surf2 = axs[1].plot_surface(X, Y, Z2, alpha=0.8, cmap='jet',
 ↪ norm=LogNorm())
axs[1].set_title("(b) Himmelblau's function (non-convex):\n"
  "$f(x,y) = (x^2 + y - 11)^2 + (x + y^2 - 7)^2$")
axs[1].set_xlabel("$x$")
axs[1].set_ylabel("$y$")
axs[1].set_zlabel("$f(x,y)$")
fig.colorbar(surf2, ax=axs[1], shrink=0.5, aspect=10,
 ↪ pad=0.15)
axs[1].view_init(elev=50, azim=-30)

# show figure
plt.show()
```

(a) Linear regression mean squared error (MSE, quadratic and convex):
$$MSE(\beta_0, \beta_1) = 0.5(2 - \beta_0 - \beta_1 0.5)^2 + 0.5(-1 - \beta_0 - \beta_1(-1.5))^2$$

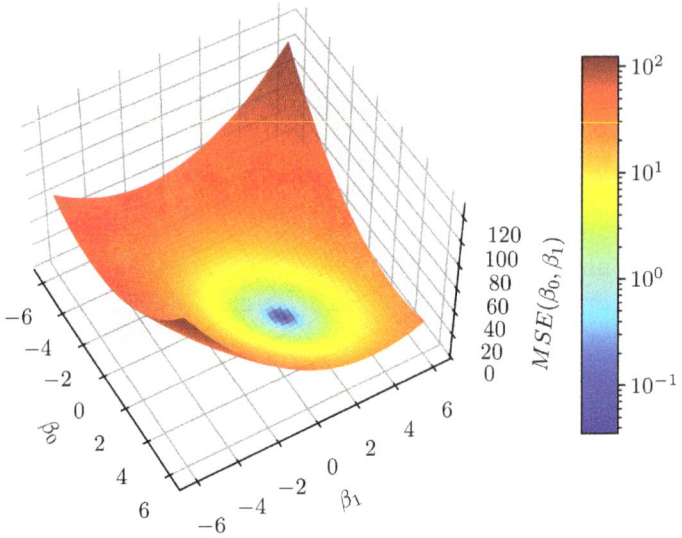

(b) Himmelblau's function (non-convex):
$$f(x, y) = (x^2 + y - 11)^2 + (x + y^2 - 7)^2$$

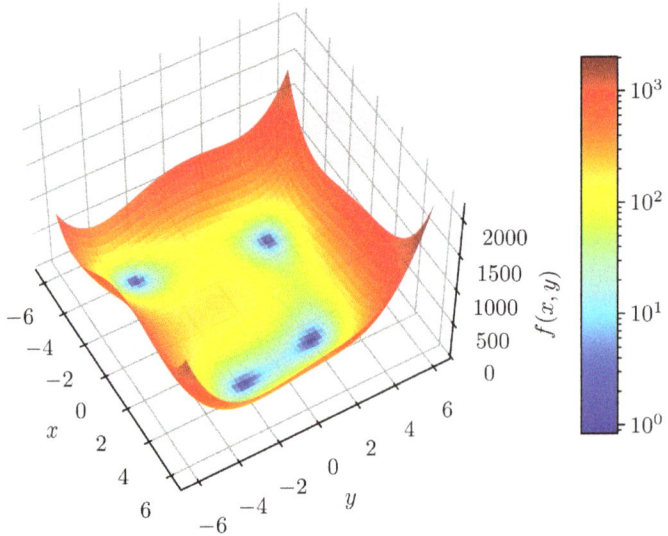

Figure 4.5: Convex vs. non-convex functions of two variables (example)

At any point on either surface in Figure 4.5, the gradient is a 2D vector that points in the direction of steepest ascent on the function surface. In the non-convex function case, you can clearly see the different minima that the gradient descent might converge to. Coincidentally, as a special feature of Himmelblau's function $f(x, y) = (x^2 + y - 11)^2 + (x + y^2 - 7)^2$, these local minima are all global minima – yielding identically low $f(x, y) = 0$. We can also see the unique minimum point for the MSE function – the optimal solution to the linear regression problem – as well as the nice convex shape of the linear regression MSE surface over the β_0 and β_1 parameter space, along which gradient descent updates can smoothly roll down towards the optimum.

4.4 Linear regression in Python code

Let us now implement the concepts we have covered using Python. We will use the well-known **diamonds** data set [ggp] from the **ggplot2** library from R programming language (also available as a `.csv` file [Wikb] from https://github.com/tidyverse/ggplot2/blob/main/data-raw/diamonds.csv; see [Lan14] for an intro to R). We will predict log10-transformed diamond prices from carat weight.

We start with the data import:

```python
import pandas as pd
data = pd.read_csv('diamonds.csv')
# alternatively, read directly from url
# data = pd.read_csv(("https://raw.githubusercontent.com/"
#
↳   "tidyverse/ggplot2/refs/heads/main/data-raw/diamonds.csv"))
data.iloc[:,:7].head()
```

	carat	cut	color	clarity	depth	table	price
0	0.23	Ideal	E	SI2	61.5	55.0	326
1	0.21	Premium	E	SI1	59.8	61.0	326
2	0.23	Good	E	VS1	56.9	65.0	327

	carat	cut	color	clarity	depth	table	price
3	0.29	Premium	I	VS2	62.4	58.0	334
4	0.31	Good	J	SI2	63.3	58.0	335

Figure 4.6 shows the relationship between the carat weight and price.

```
plt.scatter(data['carat'], np.log10(data['price']),
↪ marker="+")
plt.xlabel('Carat')
plt.ylabel('Log10 Price')
plt.show()
plt.close()
```

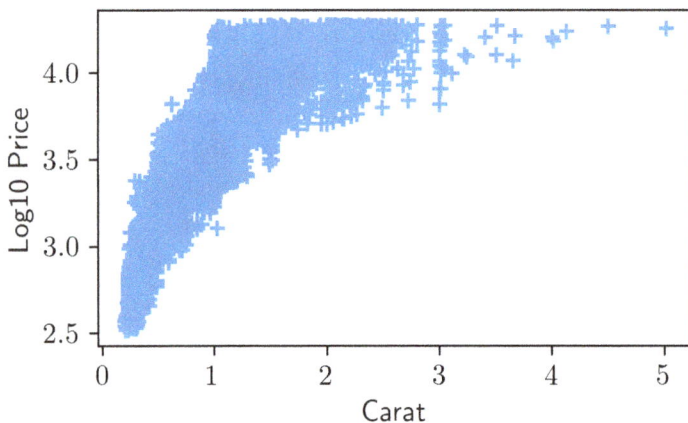

Figure 4.6: Diamonds – scatter plot

Coloring by diamond cut gives a sense of group differences:

```
# color scatter plot by cut
labels, levels = pd.factorize(data['cut'])
plt.scatter(data['carat'], np.log10(data['price']),
↪ marker="+", c=labels, cmap='tab10')
```

```
plt.xlabel('Carat')
plt.ylabel('Log10 Price')
handles = [plt.Line2D([0], [0], marker='+', linestyle='None',
 ↪  color=plt.cm.tab10(i), markersize=10) for i in
 ↪  range(len(levels))]
plt.legend(handles, levels, title='Cut', loc='lower right')
plt.show()
plt.close()
```

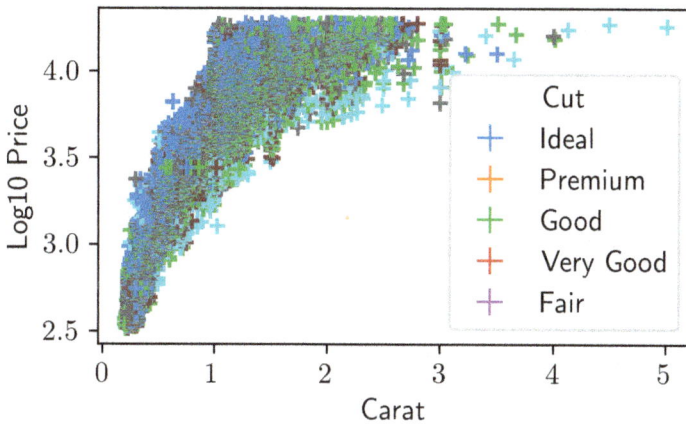

Figure 4.7: Diamonds – scatter plot by group

Preparing the data – we create an intercept column (a column of ones), to allow the model to learn a baseline offset when carat weight is zero, and transform the price:

```
data['Intercept'] = 1.
data['log10(price)'] = np.log10(data['price'])

X = data[ ['Intercept', 'carat'] ]  # input variables +
 ↪  intercept term (vector of ones)
y = data['log10(price)']  # output variable that we predict
```

4.4.1 Analytical solution (OLS)

Let us implement the closed-form OLS formula $\beta^* = (X^T X)^{-1} X^T y$:

```python
def ols(X, y):
    XtX = X.T.dot(X)
    XtXinv = np.linalg.inv(XtX)
    Xty = X.T.dot(y)
    beta = XtXinv.dot(Xty)
    return beta
```

Calculate the coefficients:

```python
beta_a = ols(X,y)
print('beta:', beta_a)
```

```
beta: [2.69914932 0.85545464]
```

The slope coefficient indicates that the logarithm with base 10 of a diamond price goes up, on average, by 0.86 with a unit increase in the carat weight of the stone. Recalling that log10 roughly gives us the number of zeros in the price, an additional carat represents a substantial value increase – in this data set.

Visualize the fitted line:

```python
plt.scatter(data['carat'], np.log10(data['price']),
    ↪  marker="+")
plt.xlabel('Carat')
plt.ylabel('Log10 Price')

x1 = np.linspace(0,2,500)
y1 = beta_a[0] + beta_a[1]*x1
plt.plot(x1, y1, "-r")
plt.xlim(0,5.5)
plt.show()
plt.close()
```

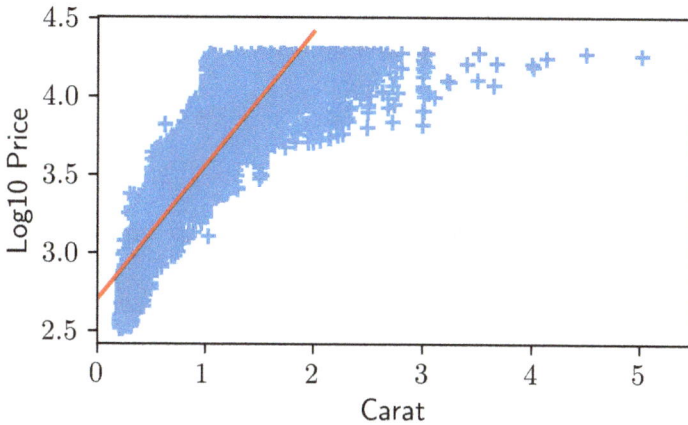

Figure 4.8: Fitted regression line

---

i Correlation vs. causation

Regression analysis results above indicate **correlation** – **not causation**. For example, we could as easily estimate a regression model of carat weight on price – and also find a positive slope coefficient – indicating that higher carat weight is associated with higher price. Clearly, however, charging a higher price does not make a diamond grow larger overnight. Interpreting regression coefficients as signifying some causal relationships is possible in some circumstances – but this requires extra assumptions. Use of regression in causal inference is a very rich area and is of great importance for analysis of experimental and observational data in medicine, basic sciences, economics, etc. – but it is outside the scope of this book. Further learning resources at the end of this chapter contain more information on this.

4.4.2 Verification with `statsmodels`

We can double-check our manual implementation using the `statsmodels` package in Python. Indeed, we get identical parameter / coefficient estimates (the `statsmodels` output contains a lot of other information, such as confidence intervals for each coefficient estimate; we do not worry about it here):

```python
import statsmodels.api as sm
model = sm.OLS(y, X).fit()
# predictions = model.predict(X)
model.summary() # identical coefficients to the analytical
 ↪  formula
# print(model.get_robustcov_results('HC3').summary()) #
 ↪  heteroscedasticity robust errors
```

Dep. Variable:	log10(price)	R-squared:	0.847
Model:	OLS	Adj. R-squared:	0.847
Method:	Least Squares	F-statistic:	2.981e+05
Date:	Tue, 09 Sep 2025	Prob (F-statistic):	0.00
Time:	17:12:56	Log-Likelihood:	18259.
No. Observations:	53940	AIC:	-3.651e+04
Df Residuals:	53938	BIC:	-3.650e+04
Df Model:	1		
Covariance Type:	nonrobust		

	coef	std err	t	P> |t|	[0.025	0.975]
Intercept	2.6991	0.001	1856.116	0.000	2.696	2.702
carat	0.8555	0.002	545.978	0.000	0.852	0.859

Omnibus:	10805.529	Durbin-Watson:	0.976
Prob(Omnibus):	0.000	Jarque-Bera (JB):	71366.235
Skew:	-0.804	Prob(JB):	0.00
Kurtosis:	8.401	Cond. No.	3.65

Notes:

[1] Standard Errors assume that the covariance matrix of the errors is correctly specified.

4.4.3 Sidebar: Python objects and classes

So far, we have used standalone functions to perform tasks like fitting a regression model. But, in practice, it is often cleaner and more scalable to organize related data and functionality using *objects* – especially when building reusable tools or libraries.

In Python, objects are defined using the `class` keyword. A class acts as a blueprint for creating objects that encapsulate both data (stored in attributes) and behavior (defined by functions). This structure improves code readability, reuse, and maintainability – especially for more complex tasks like model training and evaluation. Once a class is defined, we can create instances of that class – specific objects that hold their own data but share the same functions.

When you create an object from a class, Python automatically calls a special function named `__init__`. This method is the constructor – it sets up the object with any initial data. The first argument to all functions in a class (including `__init__`) is typically named `self`, which refers to the specific instance of the object being created or used. Using `self`, we can store values inside the object and access them later in other methods.

For example, below is a simple class that implements a linear regression. It stores the training data when initialized, fits a model by calculating regression coefficients, and includes methods to predict new values and compute the mean squared error (MSE) of the predictions. Additionally, it includes a calculation of the R^2 (R-squared) score – the proportion of variance in the target variable y that is explained by the model – defined as $R^2 = 1 - \frac{\text{MSE}}{\text{Var}(y)}$.

```python
class LinearRegression:
    def __init__(self, X, y):
        self.X = X
        self.y = y

    def ols(self, X, y):
        XtX = X.T.dot(X)
        XtXinv = np.linalg.inv(XtX)
        Xty = X.T.dot(y)
        beta = XtXinv.dot(Xty)
        return beta

    def fit(self):
        self.beta = self.ols(self.X, self.y)
        return self

    def predict(self, X_new):
        return X_new.dot(self.beta)
```

```python
    def mse(self, X, y):
        return np.mean((y-X.dot(self.beta))**2)

    def r2_score(self, X, y):
        y_pred = self.predict(X)
        # output variance - "overall" variance
        y_var = np.mean((y - np.mean(y))**2)
        # MSE - leftover unexplained variance
        mse = np.mean((y - y_pred)**2)
        # R^2 - explained variance proportion
        r2 = 1.0 - mse / y_var
        return r2
```

We can now create a `LinearRegression` object, fit it to some data, and access its coefficients and the MSE value:

```python
reg = LinearRegression(X, y).fit()

print("Coefficients:", reg.beta)
print("MSE:", reg.mse(X,y))
print("R^2 Score:", reg.r2_score(X, y))
```

```
Coefficients: [2.69914932 0.85545464]
MSE: 0.02975140412268578
R^2 Score: 0.8467801830517971
```

This object-oriented approach bundles all relevant functionality – training, predicting, and evaluating – into a single reusable entity.

Python itself is built around objects. Strings, lists, functions, and even modules are all implemented as objects under the hood. For instance, a string like "hello" is not just raw text – it is a `str` object with methods like `.upper()` or `.replace()` built in. Likewise, when we import libraries in Python (like NumPy, Pandas, or Matplotlib), we are gaining access to new objects and classes that those libraries define – objects we can create, manipulate, and use just like our own.

Understanding objects helps you get much more out of Python – and lets you build tools that are clean, reusable, and easy for others (and future-you!) to work with.

Additionally, while it is out of scope for us, you might be interested to explore how to package Python code projects and create libraries: https://packaging.python.org/en/latest/tutorials/packaging-projects/.

4.4.4 Convex optimization

As noted earlier, linear regression with the quadratic MSE objective function represents a "nice" convex optimization problem – so we can use specialized CVXPY library to solve it fast and robustly (install it via `pip install cvxpy`). We only need to provide the function of interest and initialize the parameter vector to the right shape, but do not need to compute the gradient information ourselves.

```python
import cvxpy as cp

# initialize parameter vector of length two
beta = cp.Variable(2)

# objective (minimize mean squared error)
objective = cp.Minimize(cp.sum_squares(y.values - X.values @
↪   beta) / X.shape[0])

# solve the problem
problem = cp.Problem(objective)
problem.solve()

# extract optimal value
beta_optim = beta.value

print('beta:', beta_optim)
print("MSE:", problem.value)


beta: [2.69914931 0.85545465]
MSE: 0.029751404122685798
```

Using CVXPY, we could also incorporate some constraints – for example, we could restrict intercept to be less than 2.5 (zero-weight diamond has a log10 price of less or equal to 2.5 – vs. optimal ~ 2.699):

```
# non-negativity constraint
constraints = [beta[0] <= 2.5]

# solve the constrained problem
problem = cp.Problem(objective, constraints)
problem.solve()

# extract optimal value
beta_optim = beta.value

print('beta (constrained):', beta_optim)
print("MSE (constrained):", problem.value)
```

```
beta (constrained): [2.5        1.0399345]
MSE (constrained): 0.040096318312309094
```

In both cases, because the optimization remains convex, the result we get is guaranteed to be optimal (within some small margin of error). Constrained problem MSE is higher because we prevent parameters from reaching more MSE-minimizing values via the constraint.

Note that the above CVXPY code will not work for non-convex optimization problems, such as those involved in neural net training – problem convexity is strictly required for the above code to run.

4.4.5 Black-box optimization

We can also use a general-purpose optimizer that can work on convex as well as non-convex problems to search for good parameters – we only need to provide the function of interest and the initial guess for the parameter vector, but do not need to give it the gradient information ourselves:

```
# L-BFGS-B optimizer implementation
from scipy.optimize import minimize

# initial parameter guess
beta_init = np.ones(2)

# objective function to minimize
mse = lambda beta: np.mean((y-X.dot(beta))**2)

# optimization
res = minimize(mse, beta_init, method="L-BFGS-B")

# optimal parameter
beta_optim = res.x

print('beta:', beta_optim)
```

```
beta: [2.69914933 0.85545449]
```

Such black-box optimization frameworks, relying on algorithms like the derivative-based L-BFGS-B [Wikf] above, can work decently well even with non-linear non-convex objective functions – although there is no guarantee the optimization will result in the global optimum. Some black-box optimization algorithms (including L-BFGS-B) can also support optimization problems with constraints on parameter values.

4.4.6 Gradient descent

Now let us apply the basic gradient descent update rule introduced earlier to minimize the MSE:

$$\beta_{t+1} = \beta_t - \gamma \nabla \text{MSE}(\beta_t),$$

where $\nabla \text{MSE}(\beta_t) = \frac{2}{n} X^T (X\beta_t - y)$.

To start, let us visualize the quadratic MSE shape over possible β_0 parameter values (this is like looking at Figure 4.5 (a) from β_0 side):

```
beta_temp = np.copy(beta_a)
b0 = np.linspace(0,5,100)
mse_curve = []
for b0i in b0:
    beta_temp[0] = b0i
    mse_curve.append(mse(beta_temp))
plt.plot(b0,mse_curve)
plt.xlabel('beta0')
plt.ylabel('MSE(beta0,beta1)')
plt.show()
```

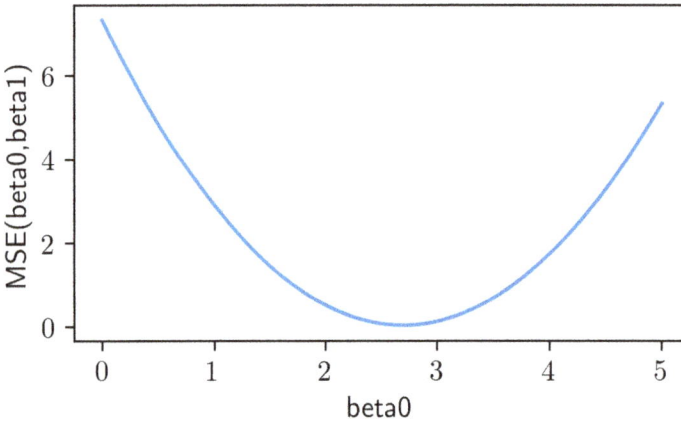

Figure 4.9: MSE shape over β_0

Gradient descent iterations:

```
beta_g = np.ones(2)
gamma = 0.01
mse_loss_hist = []
params = np.zeros((10000, 2))
for i in range(10000):
    grd = X.T.dot(X.dot(beta_g) - y) / X.shape[0]
    beta_g = beta_g - gamma * grd
    mse_loss_hist.append(mse(beta_g))
```

```
    params[i] = beta_g
print('Final beta:', beta_g.values)
```

Final beta: [2.69914737 0.85545677]

Plot the loss over time:

```
plt.plot(mse_loss_hist)
plt.xlabel('Iteration')
plt.ylabel('MSE')
plt.show()
```

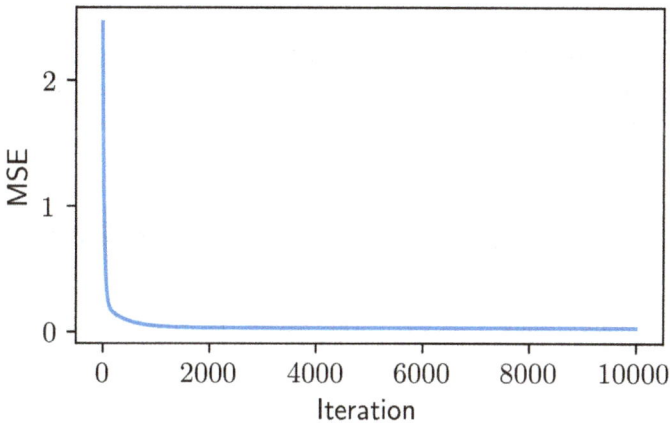

Figure 4.10: MSE history over iterations

Visualize the parameter path (color indicates the iteration number):

```
plt.scatter(params[:,0], params[:,1], c=range(10000))
plt.xlabel('beta0')
plt.ylabel('beta1')
plt.colorbar(label="Iteration")
plt.show()
```

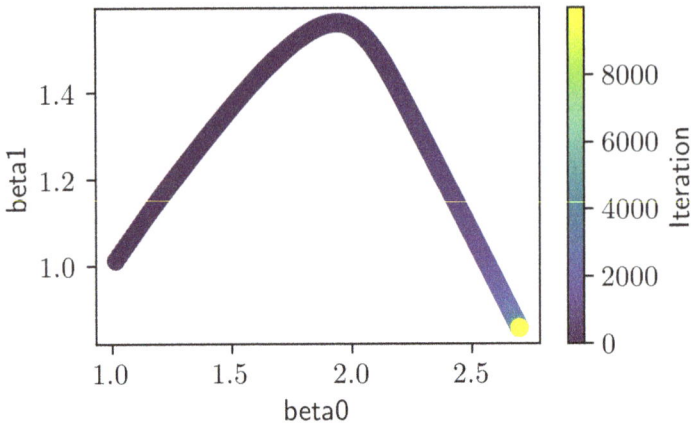

Figure 4.11: Parameter evolution over iterations

As a reminder, this gradient descent rule could be applied to minimize non-convex functions too. However, there is no guarantee it would find a global minimum of the optimized function in that case.

4.5 Exercise: Numerical gradient approximation

We have seen that MSE is a function of the parameter vector β and that we can compute the gradient analytically:

$$\nabla_\beta \text{MSE}(\beta) = \frac{2}{n} X^T (X\beta - y).$$

This gradient is crucial for optimization via gradient descent, where we repeatedly update parameters using:

$$\beta_{t+1} = \beta_t - \gamma \nabla \text{MSE}(\beta_t).$$

But what if the gradient is too complex to derive – or we simply do not want to compute it by hand? In such cases, we can *approximate the gradient numerically*, using finite differences.

Core idea:

Let β be a vector of k parameters, and let $\beta[j]$ denote its j-th component. We can write:

$$\nabla_{\beta}\text{MSE} = \begin{bmatrix} \frac{\partial \text{MSE}}{\partial \beta[1]} \\ \frac{\partial \text{MSE}}{\partial \beta[2]} \\ \dots \\ \frac{\partial \text{MSE}}{\partial \beta[k]} \end{bmatrix}.$$

Each partial derivative can be approximated, from their definition, as:

$$\frac{\partial \text{MSE}}{\partial \beta[j]} \approx \frac{\text{MSE}(\beta[0], \beta[1], \dots, \beta[j] + \Delta, \dots) - \text{MSE}(\beta[0], \beta[1], \dots, \beta[j] - \Delta, \dots)}{2\Delta}$$

for Δ small. This uses a central difference formula, where only the j-th element of β is perturbed, and the others are held constant.

Consider an example for $\beta \in \mathbb{R}^2$ (this means β has two real number elements, each can vary from negative to positive infinity).

If $\beta = \begin{bmatrix} \beta[0] \\ \beta[1] \end{bmatrix}$ then:

$$\frac{\partial \text{MSE}}{\partial \beta[0]} \approx \frac{\text{MSE}(\beta[0] + \Delta, \beta[1]) - \text{MSE}(\beta[0] - \Delta, \beta[1])}{2\Delta}$$

and

$$\frac{\partial \text{MSE}}{\partial \beta[1]} \approx \frac{\text{MSE}(\beta[0], \beta[1] + \Delta) - \text{MSE}(\beta[0], \beta[1] - \Delta)}{2\Delta}.$$

Your task:

Use this idea to implement gradient descent without the analytical gradient – just using the finite difference approximation for $\nabla_{\beta}\text{MSE}$ (gradient below). Refer to [Wikg] for details.

Solution:

```
# loading data - do not change this
import pandas as pd
data = pd.read_csv('diamonds.csv')
data['Intercept'] = 1.0
data['log10(price)'] = np.log10(data['price'])
X = data[['Intercept', 'carat']]
y = data['log10(price)']
```

```
# mse function
def mse(beta):
    return np.mean((y - X.dot(beta))**2)

gamma = 0.01 # step size
k = 2 # number of parameters
beta = np.ones(k) # initial parameter values

# gradient descent loop
for i in range(2000):
    gradient = np.zeros(k)
    # we need to fill out this vector grd with numerically
    ↪  approximated gradient

    for j in range(k):
        h = np.zeros(k)
        h[j] = 0.01
        gradient[j] = (mse(beta + h) - mse(beta - h))/0.02

    # parameter update using gradient descent - do not change
    ↪  code below
    beta = beta - gamma * gradient

print("beta", beta)
```

```
beta [2.69445593 0.86057335]
```

```
# analytical solution - to compare against it
np.linalg.inv(X.T.dot(X)).dot(X.T.dot(y))
```

```
array([2.69914932, 0.85545464])
```

The results are very close.

4.6 Multinomial logistic regression

So far, we have derived linear regression as a method to predict a continuous target variable y using input data X through a linear transformation: $X\beta$.

However, in many real-world applications, the output is **categorical**, not continuous. For example:

- Predicting which product a customer will buy.
- Predicting which class an item belongs to.
- Predicting which letter comes next in a sentence.
- Deciding between "yes" or "no" in binary classification.

These tasks fall under *logistic regression*:

- *Binary logistic regression* when there are two outcomes;
- A generalization, *multinomial logistic regression*, when there are multiple choices.

The central idea is to define a model that outputs predicted *probabilities* for each possible class – and then adjust model parameters so these probabilities align with the observed outcomes.

Here is how the multinomial logistic regression model works:

- **Utility assignment.** Each option j among J choices is assigned a utility score $u_j = x_j^T \beta$, where x_j is a vector of inputs describing option j, and β is a shared parameter vector (common across all options) – for example, β might include a negative weight for price of items. Utilities can also be called *rewards* or *logits*. They represent numerical scores that are converted into probabilities via softmax – see the next step.

- **Probability transformation (softmax).** The utilities are converted into probabilities using the softmax function:

$$P(j) = \frac{\exp(u_j)}{\sum_{k=1}^{J} \exp(u_k)}.$$

This ensures:

- Probabilities are always positive.

– Probabilities across all J options sum to 1.

Numerical trick: To control the utility magnitudes and improve the numerical stability, subtract the maximum utility $u_{\max}$ from all utilities before exponentiating:

$$P(j) = \frac{\exp(u_j - u_{\max})}{\sum_k \exp(u_k - u_{\max})}.$$

This is equivalent to dividing both the numerator and the denominator of the probability fraction by $\exp(u_{\max})$, so the probability value remains unaltered. For example,

$$\frac{\exp(u_j)}{\exp(u_{\max})} = \frac{e^{u_j}}{e^{u_{\max}}} = e^{u_j} \cdot e^{-u_{\max}} = e^{u_j - u_{\max}} = \exp(u_j - u_{\max}).$$

- **Loss function (cross-entropy).** To train the model, we compare predicted probabilities to actual outcomes using **cross-entropy** (CE) loss:

$$\mathrm{CE} = -\sum_{j=1}^{J} y_j \log P(j),$$

where:

– $y_j \in \{0,1\}$ is an indicator: 1 if option j was chosen, 0 otherwise.
– $P(j)$ is the model's predicted probability for option j.

For a single observation where class m was chosen, this simplifies to:

$$\mathrm{CE} = -\log P(m).$$

Thus, cross-entropy is just the negative log-probability assigned to the correct class. The model is trained by minimizing this loss, averaged across many choice instances.

- **Parameter estimation.** Parameters β are adjusted to minimize the cross-entropy using **gradient descent** – just like in linear regression.

- **Interpretation.** Minimizing cross-entropy is mathematically equivalent to *maximizing the likelihood* of the observed outcomes under the model – a principle known as **maximum likelihood estimation**.

> **i** Avoiding derivations by hand
>
> It is possible to derive an analytical gradient for multinomial logistic regression, but the expression becomes quite cumbersome. We will see in later chapters how **automatic differentiation** can handle gradient computation for us for use in training. For now, we could just use black box optimization or numerical gradient approximation techniques demonstrated earlier.

Softmax and cross-entropy toy example:

```python
import numpy as np

def softmax(u):
    u_max = np.max(u)  # numerical stability
    exp_u = np.exp(u - u_max)
    return exp_u / np.sum(exp_u)

# example: utility scores for 3 choices
u = np.array([1.2, -0.7, 2.5])
probs = softmax(u)
print("Softmax probabilities:", probs)

y = 0  # true class index (first option was selected)
ce = -np.log(probs[y])  # cross-entropy loss = -log predicted
↪    prob of true class
print("Cross-entropy:", ce)
```

```
Softmax probabilities: [0.20751773 0.03103814 0.76144413]
Cross-entropy: 1.5725384826797602
```

For an in-depth discussion of the multinomial logistic regression models and the related field of discrete choice modeling, I recommend [Tra09].

We will see an example of estimating a discrete choice model later in the chapter.

4.7 Minimizing error vs. maximizing likelihood

We have focused so far on optimizing model parameters by minimizing prediction error. For a linear regression, the prediction model is $y \approx \hat{y}_i = x_i^T \beta$, where x_i is vector of input variables for some observation together with a scalar 1 for intercept – and β is a vector of parameters. We could re-write this as $y_i = x_i^T \beta + \epsilon_i$, where ϵ_i is the residual or error term: $\epsilon_i = y_i - x_i^T \beta$.

Mean squared error (MSE) objective function of β, which we minimize by tweaking β, is then:

$$MSE(\beta) = \frac{1}{n} \sum_{i=1}^{n} \epsilon_i^2 = \frac{1}{n} \sum_{i=1}^{n} (y_i - x_i^T \beta)^2.$$

This is an intuitive and concrete approach – find parameters that make predicted values close to actual ones.

But there is another, probabilistic lens for formulating the optimization problem, called **maximum likelihood estimation (MLE)**.

Suppose we *assume* that the errors ϵ_i follow a Normal distribution with mean zero and variance σ^2: $\epsilon \sim N(0, \sigma^2)$. Then the outcomes $y_i = x_i^T \beta + \epsilon_i$ follow: $y_i \sim N(x_i^T \beta, \sigma^2)$.

Each observation's likelihood (probability density) is

$$p(y_i) = \frac{1}{\sqrt{2\pi\sigma^2}} \exp\left(-\frac{(y_i - x_i\beta)^2}{2\sigma^2}\right).$$

Assuming independent observations, the joint probability distribution (likelihood) of the data is:

$$L = p(y1, y2, \ldots, y_n)$$
$$= p(y_1)p(y_2|y_1)p(y_3|y2, y1) \cdots p(y_n|y_{n-1}, \ldots)$$

(decomposing into conditional densities)

$$= p(y_1)p(y_2) \cdots p(y_n) = \prod_{i=1}^{n} p(y_i)$$

(because observations in the data are considered independent)

$$= \prod_{i=1}^{n} \frac{1}{\sqrt{2\pi\sigma^2}} \exp\left(-\frac{(y_i - x_i^T \beta)^2}{2\sigma^2}\right).$$

In MLE, this product of densities is what we want to maximize by tweaking β. However, it is inconvenient to work with products of small numbers like that. A trick is to take a logarithm (usually, natural logarithm) of the product – which converts it into a sum (by property of logarithm). Logarithm is monotonously increasing function, so whatever parameters give maximum of the original function, they also give max of the log-transformed function – the log-likelihood.

$$
\begin{aligned}
LL(\beta) &= \log p(y1, y2, \ldots, y_n) \\
&= \log \prod_{i=1}^{n} \frac{1}{\sqrt{2\pi\sigma^2}} \exp\left(-\frac{(y_i - x_i^T\beta)^2}{2\sigma^2}\right) \\
&= \sum_{i=1}^{n} \log \frac{1}{\sqrt{2\pi\sigma^2}} \exp\left(-\frac{(y_i - x_i^T\beta)^2}{2\sigma^2}\right) \\
&= \sum_{i=1}^{n} \left[\underbrace{\log 1}_{=0} - 0.5\log(2\pi\sigma^2) - \frac{(y_i - x_i^T\beta)^2}{2\sigma^2}\right] \\
&= -0.5 \sum_{i=1}^{n} \left[\log(2\pi\sigma^2) + \frac{1}{\sigma^2}(y_i - x_i^T\beta)^2\right] \\
&= -0.5n \log(2\pi\sigma^2) - 0.5 \sum_{i=1}^{n} \left[\frac{1}{\sigma^2}(y_i - x_i^T\beta)^2\right],
\end{aligned}
$$

so

$$
LL(\beta) = -0.5n \log(2\pi\sigma^2) - 0.5 \sum_{i=1}^{n} \left[\frac{1}{\sigma^2}(y_i - x_i^T\beta)^2\right].
$$

This is the probability-theory-motivated objective to maximize by tweaking β. Notice that $-0.5n \log(2\pi\sigma^2)$ part does not change depending on β and so does not affect optimization and can be ignored. And the second term $-0.5 \sum_{i=1}^{n} [\frac{1}{\sigma^2}(y_i - x_i^T\beta)^2]$ is proportional to negative MSE. Maximizing the log-likelihood with respect to β is thus equivalent to minimizing MSE – under the assumption of normally distributed errors.

This duality is important:

- From a prediction perspective, we minimize error (e.g., MSE or cross-entropy).
- From a statistical perspective, we maximize the likelihood of the data.

In fact, the cross-entropy loss used in classification models also stems from a likelihood-based assumption. If we assume that each class label is drawn from a categorical distribution parameterized (represented) by the softmax-predicted probabilities, then the negative log-likelihood corresponds exactly to the cross-entropy loss. As we have already seen, the cross-entropy error reduces to the negative log-probability of the correct class, highlighting this connection.

If you are interested in learning more about the statistical perspective on regression, I recommend [JW07].

4.8 Application: Tournament evaluation of LLMs with bootstrap

Large language models (LLMs) are powerful models that can generate human-like text. We will investigate how they work later in the book.

One approach to assessing the quality of different LLMs is to ask humans which LLM outputs they prefer – as a form of crowdsourcing. A common evaluation setup presents a human rater with a pair of responses to a query – each generated by a different LLM – and asks the human to select the preferred response. Each such comparison yields a single observation indicating, at a minimum: the IDs of the compared models and the ID of the winning model (or a flag indicating a tie). The data set typically includes choices made by multiple human raters across various model pairs and queries.

The goal is to use this binary choice data to estimate a ranking of LLMs by quality, so that higher-performing models appear at the top.

A naive approach might be to compute a model's win rate against each competitor separately and then take an average. However, this requires complete pairwise comparison data, with multiple comparisons per each model pair, which is rarely available. Another approach is to count how many times a model is preferred over all others, normalized by the total number of comparisons. But this too can be misleading – a model may achieve a deceptively high rating simply because it is frequently compared against weaker models. Furthermore, if models are closely matched in quality, the probability of any one model winning decreases as the number of competitors increases. This can introduce systematic bias in win ratios that reflects the evaluation setup more than the models' intrinsic quality.

These issues can be addressed by fitting a simple logistic regression model to predict which LLM wins in a paired comparison, given the identities of the competing models. The resulting parameter estimates, called utilities (alternatively, rewards or logits), reflect model quality, as determined by the comparisons and preferences that generated them, and can be used to produce a ranking of LLMs.

This approach is known as a Bradley-Terry model [Wika] and is related to the Elo rating system [Wikd] used in chess and esports. The Bradley-Terry model has been adopted for LLM evaluation using human preference data – for example, in [Chi+24; Chi+25; Zhe+23]. This approach is also called *tournament evaluation*. We will use the data set from [Chi+24] released on Hugging Face (data was collected between June and August 2024): https://huggingface.co/datasets/lmarena-ai/arena-human-preference-100k.

The data set comes in `.parquet` file format, used to store large data sets – Pandas can read it in:

```
import pandas as pd
data = pd.read_parquet(
    "./arena-explorer-preference-100k.parquet")

# alternatively, read directly from the url
# url = ("https://huggingface.co/datasets/lmarena-ai/"
#         "arena-human-preference-100k/resolve/main/"
#         "data/arena-explorer-preference-100k.parquet"
#         )
# data = pd.read_parquet(url)

data = data[['model_a', 'model_b', 'winner']] # keeping just
    three columns
data.head()
```

	model_a	model_b	winner
0	claude-3-5-sonnet-20240620	gpt-3.5-turbo-0125	tie (bothbad)
1	mistral-large-2407	athene-70b-0725	model_b
2	claude-3-opus-20240229	gemini-1.5-flash-api-0514	tie (bothbad)

	model_a	model_b	winner
3	gemma-2-9b-it	qwen2-72b-instruct	model_b
4	mixtral-8x22b-instruct-v0.1	llama-3.1-70b-instruct	tie (bothbad)

How a winner in a paired comparison is recorded:

```
data['winner'].unique().tolist()
```

```
['tie (bothbad)', 'model_b', 'model_a', 'tie']
```

For simplicity, we will filter out data with ties:

```
# remove ties
data = data[~data['winner'].str.contains('tie'
    ↪ )].reset_index(drop=True)
data['winner'].unique().tolist()
```

```
['model_b', 'model_a']
```

Distinct LLM models appearing in the data, in the alphabetic order:

```
model_names = sorted(data[['model_a',
    ↪ 'model_b']].stack().unique().tolist())
```

A basic data set description:

```
n = data.shape[0]
m = len(model_names)

print("Number of comparisons:", n)
print("Number of LLMs:", m)
```

```
Number of comparisons: 65418
Number of LLMs: 55
```

We will associate a utility score u_j with each of m evaluated LLMs, $j \in 1 : m$. We will store these utility scores in a vector u of length m. As usual, we do not know, what the utility scores should be a priori – we need to estimate u from the data.

We will write the probability that a model a is preferred over a model b – in an a vs. b comparison – as:

$$P(a \succ b) = \frac{\exp(u_a)}{\exp(u_a) + \exp(u_b)}$$
$$= \frac{\exp(u_a - u_b)}{\exp(u_a - u_b) + 1}$$
$$= \frac{1}{1 + \exp(-(u_a - u_b))},$$

where we get the last two equations by dividing the first expression's numerator and denominator by either $\exp(u_b)$, or $\exp(u_a)$ – the probability value remains unchanged. Then, the log-probability is

$$\log P(a \succ b) = \log \frac{1}{1 + \exp(-(u_a - u_b))} = -\log(1 + \exp(-(u_a - u_b))).$$

If we kept ties, these could be handled via a generalized Bradley-Terry model with ties – but that adds too much extra complexity for our purposes.

We interpret these utilities as measures of LLM quality – because the higher an LLM's estimated utility, the higher is its chance of beating another model in a paired comparison. Sometimes these utility parameters are referred to as *latent variables* – because they are not directly observed, but are inferred indirectly through their effect on binary comparison outcomes.

The log-likelihood of ith comparison, as a function of the model quality vector u, is the log-probability:

$$LL_i(u) = \log P(\text{winner (i)} \succ \text{loser (i)}).$$

The log-likelihood of the full data set is the sum of $LL_i(u)$ across all n comparisons in the data:

$$LL(u) = \sum_{i=1}^{n} \log P(\text{winner (i)} \succ \text{loser (i)}).$$

4.8.1 Identification and regularization

We could already attempt to maximize this log-likelihood by optimizing the vector of LLM utility scores u. However, a conceptual issue remains. Unlike a linear regression, where a unique combination of parameters exists that minimizes the MSE error (as long as $X^T X$ is invertible and the OLS solution exists), in the case of the log-likelihood objective for paired comparisons, multiple combinations of utility scores – the parameters we need to estimate – can yield the same exact optimal log-likelihood value. This phenomenon is known as *non-identifiability*.

To see why, consider a paired comparison probability:

$$P(a \succ b) = \frac{1}{1 + \exp(-(u_a - u_b))}.$$

This probability depends *only on the difference* $\Delta = u_a - u_b$. That is,

$$P(a \succ b) = \frac{1}{1 + \exp(-\Delta)}.$$

Now observe that adding a constant to both u_a and u_b does not change Δ, and thus has no effect on the likelihood. For example, $\Delta = 100 - 99 = 1 - 0 = 1$: whether $u_a = 100$ and $u_b = 99$ or $u_a = 1$ and $u_b = 0$, the probability remains the same. Hence, the absolute scale of the utility values is arbitrary – the model is *invariant* to translation (shift) of all parameters. The paired comparison data can be equally well described by infinitely many equally good u solutions differing only in this shift.

This may not seem like a problem. If the optimization procedure finds some utility vector that maximizes the log-likelihood, why worry about alternatives that would produce the same ranking? However, in practice, this can cause instability during optimization. If the values of u drift to large magnitudes, gradient steps can become too aggressive or noisy, slowing or destabilizing convergence.

A straightforward solution is to add to the objective function a *regularization* term that penalizes large parameter magnitudes. For example, we can add an l_2 (ridge) regularization term – the sum of squared utilities, which has to be minimized during training, concurrently with the primary / focal objective function:

$$\sum_{j=1}^{m} u_j^2.$$

This encourages the optimizer to anchor the utilities near zero, resolving the scale ambiguity and making the problem identifiable in practice. Regularization can be useful in a variety of modeling settings, including in linear regression estimation, where it makes the linear regression parameters less sensitive to possible noise / outliers and also ensures the existence of a unique optimal β solution when $X^T X$ is not invertible. Regularization strength can be varied by pre-multiplying the regularization objective by a positive constant $\lambda > 0$ (pronounced *lambda*): $\lambda \sum_{j=1}^{m} u_j^2$. A suitable value for λ is usually chosen by trying different values and seeing which value yields the model that performs best on held-out data not used in training. Here, we will set $\lambda = 1$ for simplicity.

Another issue, related to identification, that may arise in paired comparison data has to do with data coverage of possible model pairs. Consider having four LLMs A, B, C, and D.

- If the data includes no direct comparisons between A and B, but both have been compared to C and D (and C and D to each other), then we can still infer indirect rankings: e.g., A is ranked above C, B is below C, so A should be above B.
- But if comparisons occur only within the (A, B) pair and within the (C, D) pair – and none across the two groups – we cannot determine how A / B compares to C / D. In this case, even though adding regularization still ensures convergence, utilities between the two LLM groups are not comparable and the resulting ranking would not be valid.
- Even more subtly, if only A is compared to other models (B, C, D), and A is always preferred, we can say A is likely better – but we learn nothing about how B, C, and D compare to each other.

Thus, it is crucial in practice to cover by observed comparisons as many possible model pairs as possible. More formally, we aim for the data to, at the very least, form a *connected comparison graph*, meaning there is a comparison path (direct or indirect) between every pair of models. In our case, a comparison graph is a network where each node represents an LLM and an edge between two nodes means that the two LLMs have been directly compared at least once. Concepts like connectedness of a graph are studied in *graph theory* [EK+10].

Luckily, in our case, the data set is large. A quick check confirms that the comparison data covers the large majority of LLM pairs:

```
# sort LLM names within each comparison pair alphabetically
pairs = data.apply(lambda row:
  tuple(sorted([row['model_a'], row['model_b']])), axis=1)
# count the number of unique LLM pairs in the data
n_pairs = len(pairs.unique())
# divide by the number of possible unique pairs (m choose 2)
coverage = n_pairs / (m*(m-1)/2)
print('Proportion of possible LLM pairs observed:', coverage)
```

```
Proportion of possible LLM pairs observed: 0.8134680134680135
```

We can also confirm the connectedness of the comparison graph as follows:

```
import networkx as nx
G = nx.Graph()
G.add_edges_from(data[['model_a', 'model_b']].values)
is_connected = nx.is_connected(G)
print("Is the comparison graph connected?", is_connected)
```

```
Is the comparison graph connected? True
```

The detailed mathematical conditions for identification in paired comparison models are discussed in [For57; Hun04].

4.8.2 Estimating the Bradley-Terry model

To begin, we encode unique LLM names using a numeric index:

```
m2i = {m:i for i,m in enumerate(model_names)}
data['model_a_ind'] = data['model_a'].map(m2i)
data['model_b_ind'] = data['model_b'].map(m2i)
```

We then initialize the parameter vector of m real numbers $u \in \mathbb{R}^m$, where each u_j represents the utility or quality score of the jth LLM – among m total evaluated LLMs:

```
u_init = np.ones(m)
```

We aim to estimate the utility scores by fitting the Bradley-Terry model via maximum likelihood estimation, regularized to ensure identifiability and numerical stability. Specifically, we minimize the following *loss* function, consisting of the *negative* log-likelihood and the positive regularization terms:

$$\mathcal{L}(u) = -LL(u) + \sum_{j=1}^{m} u_j^2$$

$$= -\sum_{i=1}^{n} \left[-\log(1 + \exp(-(u_{\text{winner (i)}} - u_{\text{loser (i)}}))) \right]$$

$$+ \sum_{j=1}^{m} u_j^2.$$

Note that *maximizing $LL(u)$ is equivalent to minimizing its negative* $-LL(u)$.

This objective encourages correct prediction of comparison outcomes while penalizing large utility values.

Objective function via a for loop (slow, but easier to understand):

```
def nLL(u, data):
    total_log_p = 0
    for i in range(data.shape[0]): # for each comparison
        w = data['winner'].iloc[i]
        m_a = data['model_a_ind'].iloc[i]
        m_b = data['model_b_ind'].iloc[i]
```

```
        if w == 'model_a':
            delta = u[m_a] - u[m_b]
        else:
            delta = u[m_b] - u[m_a]
        total_log_p += -np.log(1 + np.exp(-delta))
    return -total_log_p + np.sum(u**2) # -LL + regularization
  ↪
```

```
nLL(u_init, data)
```

```
np.float64(45399.30225780515)
```

The function above turns out to be slow. In fact, we can avoid using a for loop by taking advantage of clever vectorization:

```
def nLLv(u, data):
    # difference u[m_a] - u[m_b] across rows
    delta = u[data['model_a_ind'].values] -
  ↪ u[data['model_b_ind'].values]
    # invert the sign if b won over a in a given comparison
    delta *= 2 * (data['winner'] == 'model_a') - 1
    # as before
    log_p = - np.log(1 + np.exp(-delta))
    return -np.sum(log_p) + np.sum(u**2)
```

```
nLLv(u_init, data)
```

```
np.float64(45399.3022578705)
```

This function computes the loss in a fully vectorized way, making it suitable for optimization over large-scale datasets.

We can now find the LLM utilities that best fit the data using black-box optimization:

```
from scipy.optimize import minimize
import time

start_time = time.time()
res = minimize(nLLv, u_init, args=(data), method="L-BFGS-B",
    ↪  tol=0.00001)
exec_time = time.time() - start_time

print("Finished in %s seconds" % (round(exec_time,2)))
print("First 4 u_j values:", res.x[:4])
print("Loss (-LL+reg):", res.fun)
```

```
Finished in 9.15 seconds
First 4 u_j values: [ 0.57600542  1.17221523  0.73590724 -
0.1414109 ]
Loss (-LL+reg): 41913.59468584251
```

Convexity. This optimization problem, coincidentally, turns out to be convex. The objective function we minimize, after simplification, is

$$\mathcal{L}(u) = \sum_{i=1}^{n} \log(1 + \exp(-(u_{\text{winner (i)}} - u_{\text{loser (i)}}))) + \sum_{j=1}^{m} u_j^2$$

and represents the sum of:

- positive log-sum-exp components: $\log(1+\exp(-(u_{\text{winner (i)}} - u_{\text{loser (i)}})))$;
- and positive quadratic components: u_j^2.

These constituent function components (positive log-sum-exp and positive quadratic) turn out to be convex in the parameter vector u, and their sum is also convex [BV04]. We are searching for the optimal parameter vector u over the unconstrained domain $\mathbb{R}^m$ of real numbered vectors, which is a convex set. Therefore, we have a minimization of a convex function over a convex set – a convex optimization problem. This guarantees that any local minimum is a global minimum and we could have also used a convex optimization library like CVXPY here, which would allow additional constraints (e.g., bounded utilities) to be easily incorporated.

Further, the regularization quadratic function $\sum_{j=1}^{m} u_j^2$ is *strictly* convex in vector u (as a quadratic function), and so it has a unique minimizer – trivially, $u = 0$. As discussed earlier, a sum of a convex function and a strictly convex function, defined on the same convex domain, must be strictly convex, so the whole function $\mathcal{L}(u)$ is necessarily strictly convex – thus, there exists a unique global minimizer u of $\mathcal{L}(u)$, rendering the problem identifiable.

Convex solution (we have to rewrite parts of the objective function so that CVXPY can track the arithmetic operations and ensure the convexity of the problem):

```python
import cvxpy as cp

def conv_tournament(data):
    u = cp.Variable(m) # LLM params
    delta = u[data['model_a_ind'].values] -
        u[data['model_b_ind'].values]
    delta = cp.multiply(2 * (data['winner'] == 'model_a') -
        1, delta)
    log_p = -cp.logistic(-delta)
    objective = cp.Minimize(-cp.sum(log_p) +
        cp.sum_squares(u))
    problem = cp.Problem(objective)
    problem.solve(cp.CLARABEL, max_iter=5) # specific solver
        settings, 5 iterations are enough here and save time, but
        the default 50 should be used for more general problems
    return u.value, problem.value

start_time = time.time()
u_opt, loss_opt = conv_tournament(data)
exec_time = time.time() - start_time


print("Finished in %s seconds" % (round(exec_time,2)))
print("First 4 u_j values:", u_opt[:4])
print("Loss (-LL+reg):", loss_opt)


Finished in 2.37 seconds
```

```
First 4 u_j values: [ 0.55954551  1.15973313  0.71777079 -
0.16122639]
Loss (-LL+reg): 41912.093227506164
```

In this case, using the convex framework results in a slightly more exact / optimal solution, while also getting us there faster than the L-BFGS-B optimizer. (Although this performance pattern might not generalize to problems with more parameters or different error tolerance.)

4.8.3 Bootstrap for quantifying uncertainty in the estimated parameters

Before diving into the visualization of LLM quality scores, it is important to quantify the uncertainty in these estimates.

When we estimate a model's quality u_j, we are doing so on a finite data sample that comes from some unobserved data-generating distribution. Our goal is not just to summarize the data sample itself, but to infer properties of that full distribution. Because any finite sample provides only a noisy snapshot of the underlying population, different samples will yield different estimates. Therefore, we analyze the *sampling distribution* of the parameter of interest – the theoretical distribution of the estimate across repeated random data samples from the true population. If we knew the shape of such a sampling distribution for each u_j, we could construct the confidence interval marking the plausible range for an LLM's true utility – going beyond a potentially noisy point estimate.

There are many ways to get to the shape of the sampling distribution – this is the domain of study of *statistical inference*. One of the simplest and most popular approaches is the *bootstrap*. Bootstrap works by simulating the process of drawing new datasets from the underlying data-generating distribution.

The core idea is as follows: we repeatedly *resample* the original data set of n observations *with replacement* to create many bootstrap sample data sets of n observations each. Sampling with replacement means the same observation / row in the original data can be selected multiple times, and some rows may never get selected. For each new bootstrap-sampled data set, we re-estimate the parameters of interest. This gives us a distribution of parameter estimates, from which we can calculate the confidence intervals. The presented bootstrap algorithm is called *nonparametric bootstrap*, as it relies only on resampling the observed data – without a need to specify a parametric distribution for drawing samples.

The theoretical justification for this bootstrap algorithm is that, under general conditions, the distribution of estimates computed using bootstrap samples approximates the true sampling distribution of the estimator – that is, the distribution of parameter estimates we would get if we repeatedly collected new datasets from the original population [BF81; TE93; Vaa98]. Critically, this method does not require strong assumptions about the form of the underlying sampling distribution for the estimated parameter, unlike classical methods that often assume normality.

In general, taking 1000 or even 10000 bootstrap data samples is recommended; but we will stick to 100 to save time. With 100 bootstrap samples, we would get 100 u_j estimates for each of m LLMs, $j \in 1 : m$. This set of utility scores should be distributed similarly to what we would observe if we collected 100 new paired comparison human preference data sets of the same size n (which would be expensive!) and computed u_j scores using those. Instead, we can estimate the center and the confidence interval of this sampling distribution by, for example, computing the median and the 2.5th and 97.5th percentiles across the 100 bootstrap estimates of the u_j score.

Here is how it works in practice:

```
# bootstrap
np.random.seed(999)

start_time = time.time()
results = []
# number of bootstrap samples (1000+ is often recommended)
for i in range(100):
    d = data.sample(data.shape[0], replace=True)
    u_opt, _ = conv_tournament(d) # optimal u on d
    results.append(u_opt)

results = np.vstack(results) # to array
exec_time = time.time() - start_time
```

```
print("Finished in %s minutes" % (round(exec_time/60,2)))
print("(num. samples, num. LLMs):", results.shape)
```

```
Finished in 3.13 minutes
(num. samples, num. LLMs): (100, 55)
```

This took a bit of time. Now, using the utility vectors estimated from bootstrap samples, we can, for each LLM, compute the median and 2.5th and 97.5th percentiles of u_j values:

```
# percentiles of the utility estimates per LLM across
  ↪ bootstrap samples
u_median = np.median(results, 0)
u_2_5 = np.percentile(results, 2.5, axis=0)
u_97_5 = np.percentile(results, 97.5, axis=0)
# error bars (for plotting)
lower_err = u_median - u_2_5
upper_err = u_97_5 - u_median
error_bars = np.array([lower_err, upper_err])

# plotting
sort_ind = np.argsort(u_median) # sort index
plt.figure(figsize=(6, 9))
plt.errorbar(u_median[sort_ind],
        np.array(model_names)[sort_ind],
        xerr=error_bars[:, sort_ind],
        fmt='o', color = "blue",
        ecolor="blue", capsize=3)
plt.xlabel(r'Median response quality ($u_j$) with 95\% CI')
plt.ylabel('LLM')
plt.grid(True, linestyle=':', linewidth=0.5)
plt.savefig("bootstrap.png")
plt.show()
plt.close()
```

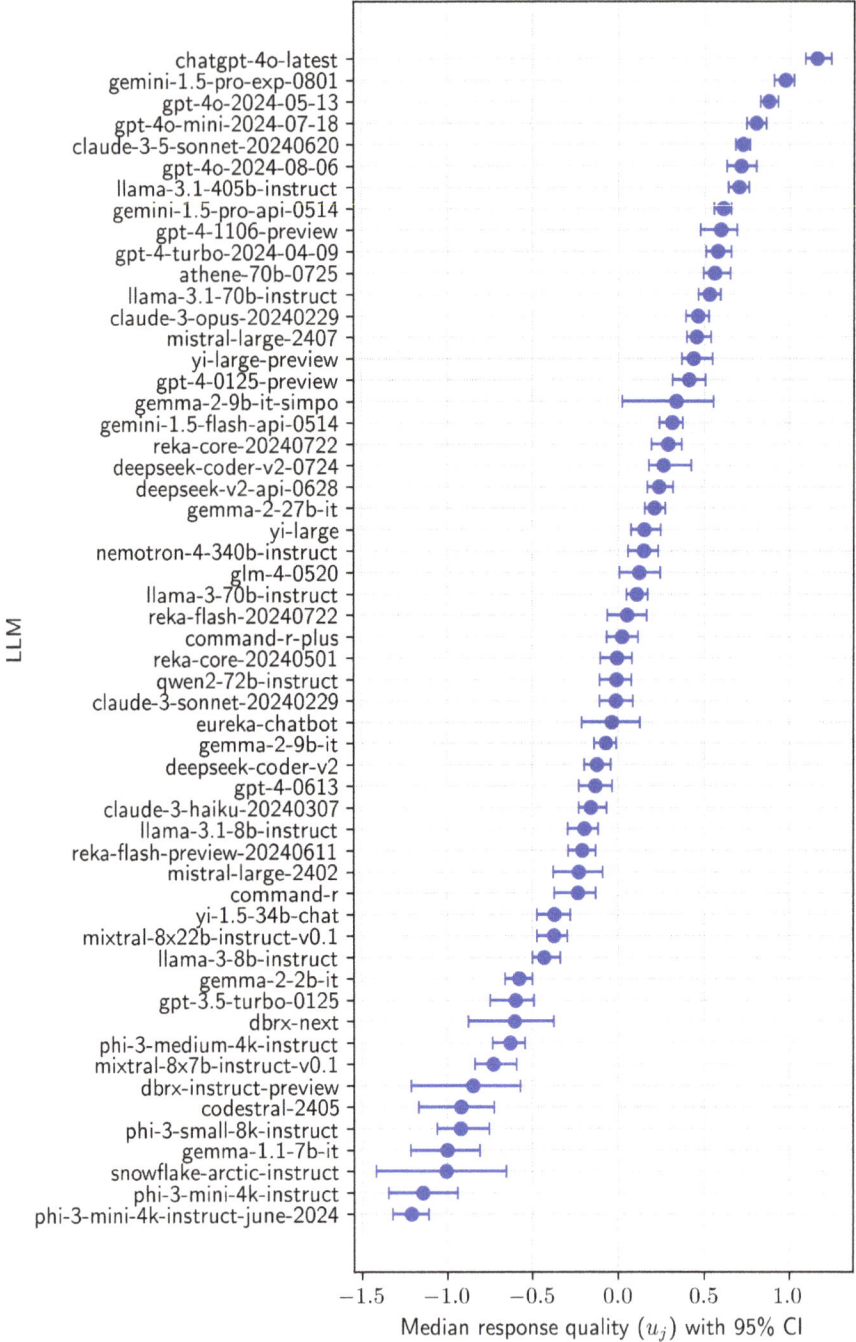

Figure 4.12: Estimated LLM quality based on human preference data from June-August 2024. Median utility with 95% confidence interval (CI: 2.5-97.5 percentile), 100 bootstrap samples.

The results in Figure 4.12 suggest that, as of August 2024, ChatGPT-4o latest model consistently outperformed other LLMs in terms of human preference.

The 95% confidence intervals (CIs) based on 2.5th and 97.5th bootstrap percentiles provide an estimate of where the true performance of each LLM likely lies, taking into account sampling variability in the data. The fact that these CIs are quite narrow and do not overlap for many of the models assures us somewhat that we would get a pretty similar ranking if we recomputed it on a different sample following the same data collection procedure within the same time period. (Contrast this with a hypothetical situation of very wide overlapping CIs – this would mean the rankings between data samples would be unstable and could change dramatically.)

We can also plot the histogram of 100 utility estimates for the LLM `chatgpt-4o-latest`:

```
plt.hist(results[:,m2i["chatgpt-4o-latest"]], bins=25)
plt.xlabel("$u_j$ score")
plt.ylabel("Frequency")
plt.show()
plt.close()
```

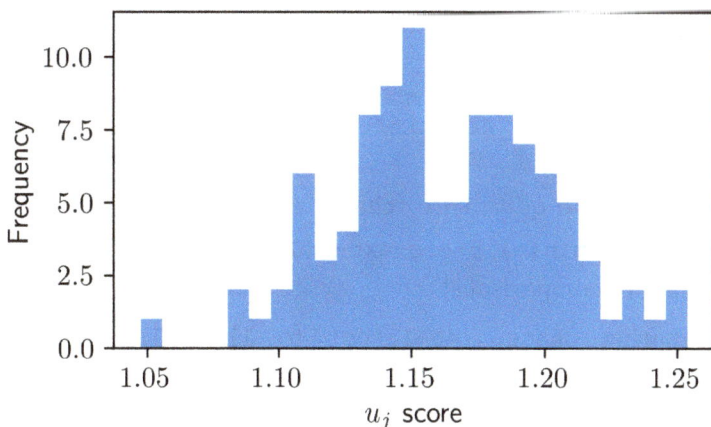

Figure 4.13: Histogram of utility scores estimated on 100 bootstrap samples for `chatgpt-4o-latest` LLM.

It is always a good idea to visualize the shape of the bootstrap distribution to better understand it beyond the percentile confidence intervals. In Figure 4.13, the shape is symmetric and normal-like, which supports the reliability of the percentile interval. In other estimation problems, we could get shapes that exhibit asymmetry (skewness), multimodality (multiple peaks), or extreme values (outliers). For example, numerous outliers might suggest that the number of bootstrap samples is too small to capture the full shape of the sampling distribution, reducing the reliability of the estimated confidence interval. In fact, Figure 4.13 does show some lone outlier values on the sides – so we would likely benefit from more bootstrap samples.

Finally, we can use the bootstrap results to get the probability that one model's output is judged to be better than another model's output in a paired comparison, using equation

$$P(a \succ b) = \frac{\exp(u_a - u_b)}{\exp(u_a - u_b) + 1}.$$

For example:

```
delta = u_median[m2i["chatgpt-4o-latest"]] -
    u_median[m2i["gemini-1.5-pro-exp-0801"]]
p = np.exp(delta) / (np.exp(delta) + 1)
print("P(chatgpt-4o-latest > gemini-1.5-pro-exp-0801) =",
  round(p,2))
```

```
P(chatgpt-4o-latest > gemini-1.5-pro-exp-0801) = 0.55
```

So, while we can be quite sure `chatgpt-4o-latest` is better than its closest follower `gemini-1.5-pro-exp-0801`, we expect it to be preferred only in $\sim 55\%$ of comparisons!

When comparing `chatgpt-4o-latest` to `llama-3.1-405b-instruct`, the seventh model in the ranking, we get:

```
delta = u_median[m2i["chatgpt-4o-latest"]] -
    u_median[m2i["llama-3.1-405b-instruct"]]
p = np.exp(delta) / (np.exp(delta) + 1)
```

```
print("P(chatgpt-4o-latest > llama-3.1-405b-instruct) =",
  round(p,2))
```

```
P(chatgpt-4o-latest > llama-3.1-405b-instruct) = 0.61
```

Users prefer `chatgpt-4o-latest` to `llama-3.1-405b-instruct` in 61% of cases – perhaps, a more tangible outperformance.

Overall, such probabilities of a paired comparison win give us a more intuitive sense of model performance than the raw utilities (which are still critical for the estimation). In fact, we could also rank LLMs by their chance of beating some benchmark model and compute the bootstrap confidence intervals for those probabilities.

While this analysis is relatively simple – we do not control for query types, annotator effects, or domain-specific biases – it is sufficient to obtain a sense of model-level performance.

4.8.4 When to use or not to use bootstrap

The bootstrap approach to uncertainty quantification is powerful and flexible. It can be applied far beyond Bradley-Terry model. For instance, it could be used to estimate parameter uncertainty in a linear regression or other more complex predictive models.

However, bootstrap is not always the best tool for the job. Some important caveats include:

- *When exact parameter uncertainty formulas exist*, as in linear regression, analytical expressions should be preferred for both speed and accuracy [JW07; Hay11; CB24]. For instance, `statsmodels` library reports analytical confidence intervals for the linear regression parameter estimates.
- *Bayesian inference* [Gel+13] is often preferable when you want to incorporate prior knowledge, quantify uncertainty more comprehensively, or handle hierarchical / complex models.

- Bootstrap fails for *extreme statistics* like maxima or minima. For example, resampling from the data cannot produce a new maximum higher than the one observed in the original data set – meaning the bootstrap distribution is biased toward the center. These cases require specialized methods such as extreme value theory [Col+01]. This is in contrast to central tendency statistics, where bootstrap works well, such as a median or a mean of a variable or regression coefficients capturing the *average* covariance between variables.
- More generally, one needs to be cautious with bootstrap when the estimate of interest is *not a smooth function of the data set* – and can jump discontinuously. For example, see the discussion of bootstrap use to estimate confidence levels for phylogenetic trees in biology [EHH96]. We will also see this issue crop up when we talk about neural nets, where small changes in the data can lead to very different estimates for a specific parameter.

That said, for many practical problems bootstrap offers a remarkably simple and effective way to quantify uncertainty – for example, when you are estimating parameters of a complex model and no closed-form uncertainty estimates exist; or if some of the distributional / CLT assumptions required for the validity of classical analytical confidence intervals do not hold.

4.9 Advanced: Analytical OLS confidence intervals

For completeness, I provide the derivation of analytical confidence intervals for the linear regression parameters. As you will see, this is a bit gnarlier than bootstrapping!

We assume the standard theoretical linear model, where β is the true unobserved vector of k linear regression coefficients that generates the data for us:

$$y = X\beta + \epsilon,$$

with $\epsilon \sim N(0, \sigma^2 I_n)$, where I_n is an $n \times n$ identity matrix; n is the number of independent identically distributed (i.i.d.) observations.

Ordinary least squares (OLS) estimator is:

$$\hat{\beta} = (X^T X)^{-1} X^T y.$$

We require here that $X^T X$ should be invertible. Substitute the true model $y = X\beta + \epsilon$ into the expression for $\hat{\beta}$:

$$\begin{aligned} \hat{\beta} &= (X^T X)^{-1} X^T y \\ &= (X^T X)^{-1} X^T (X\beta + \epsilon) \\ &= \underbrace{(X^T X)^{-1} X^T X}_{I_k} \beta + (X^T X)^{-1} X^T \epsilon \\ &= \beta + (X^T X)^{-1} X^T \epsilon. \end{aligned}$$

Note that our estimate $\hat{\beta}$ is a function of the random variable ϵ and so is a random variable itself – with an associated distribution. Also note that $A = (X^T X)^{-1} X^T$ is a constant $k \times n$ matrix.

What is the distribution of $\hat{\beta}$? Two arguments can be made:

1. If our assumption is correct, and ϵ is indeed a vector of normally distributed independent random variables, then, $(X^T X)^{-1} X^T \epsilon = A\epsilon$ is a vector of linear combinations (weighted sums) of n independent random variables (ϵ entries). By the property of a normal distribution, a linear combination of independent normal random variables $A\epsilon$ is also normally distributed.

2. If the normality assumption is violated and ϵ is not normally distributed, the quantity $A\epsilon$ – being a linear transformation of ϵ – still consists of weighted sums of the n components of the error vector. In this case, the Central Limit Theorem (CLT) provides a justification – under some mild conditions and as n becomes large, each component of $A\epsilon$ is approximately normally distributed. Even when the components of ϵ are not perfectly independent, violating the standard CLT – as is often the case in time-series data – certain generalized versions of the CLT may still apply, provided the dependence is weak enough.

Adding a constant β just shifts the center of $A\epsilon$ distribution. Therefore, in general, the distribution of $\hat{\beta} = \beta + (X^T X)^{-1} X^T \epsilon$ should be approximately normal in large enough samples, making inference (drawing conclusions from data) based on normality a reasonable approximation.

The expected value of $\hat{\beta}$ is:

$$E[\hat{\beta}] = E[\beta + (X^TX)^{-1}X^T\epsilon]$$
$$= \beta + (X^TX)^{-1}X^TE[\epsilon]$$
$$= \beta$$
$$\text{(since } E[\epsilon] = 0).$$

The distribution of $\hat{\beta}$ is centered at the true parameter β, which is encouraging.

The second moment about the center is the covariance:

$$\text{Cov}(\hat{\beta}) = E[(\hat{\beta} - \beta)(\hat{\beta} - \beta)^T]$$
$$= E[(X^TX)^{-1}X^T\epsilon \cdot \epsilon^TX(X^TX)^{-1}]$$
$$= (X^TX)^{-1}X^TE[\epsilon\epsilon^T]X(X^TX)^{-1}$$
$$\text{(assuming } E[\epsilon\epsilon^T] = \sigma^2I_n)$$
$$= \sigma^2(X^TX)^{-1}\underbrace{X^TX(X^TX)^{-1}}_{I_k}$$
$$= \sigma^2(X^TX)^{-1}.$$

And so, assuming normality holds,

$$\hat{\beta} \sim N(\beta, \sigma^2(X^TX)^{-1}).$$

---

i Constant variance assumption

In some data sets, we might doubt the accuracy of the constant variance (*homoscedasticity*) assumption $E[\epsilon\epsilon^T] = \sigma^2I_n$, required to simplify the covariance formula for $\hat{\beta}$ estimate. For example:

- If there is non-constant variance, also called *heteroscedasticity* (e.g., variance in income increases with age or education);
- If off-diagonal covariance terms are non-zero (e.g., autocorrelation in time series);
- If there are clusters / groups of observations with distinct covariance blocks.

In such cases, we should try to get a better estimate of $E[\epsilon\epsilon^T]$

---

from the data; there are many names and approaches for this, including *heteroscedasticity robust standard errors, autocorrelation-consistent standard errors,* or *clustered standard errors.* Package `statsmodels` supports many such options: https://www.statsm odels.org/dev/generated/statsmodels.regression.linear_mode l.OLSResults.get_robustcov_results.html. For example, in `statsmodels`, after running `model = sm.OLS(y, X).fit()`, we can execute `model.get_robustcov_results('HC3').summary()` for heteroscedasticity robust standard errors (in general, a pretty decent default setting).

Standard error (SE) is just a name for the standard deviation characterizing the sampling distribution of the parameter estimate $\hat{\beta}_j$. Given $\hat{\beta} \sim N(\beta, \sigma^2 (X^T X)^{-1})$, the standard error for $\hat{\beta}_j$ is:

$$SE(\hat{\beta}_j) = \sqrt{\sigma^2 (X^T X)^{-1}_{jj}},$$

where $(X^T X)^{-1}_{jj}$ means jth diagonal entry of $(X^T X)^{-1}$.

We are almost there, but we do not actually know the true σ^2, so we estimate it from data. Recalling $\hat{y} = X\hat{\beta}$, we get $\hat{\epsilon} = y - \hat{y}$. The unbiased estimator of σ^2 is

$$\hat{\sigma}^2 = \frac{1}{n-k} \hat{\epsilon}^T \hat{\epsilon}.$$

We get

$$SE(\hat{\beta}_j) = \sqrt{\hat{\sigma}^2 (X^T X)^{-1}_{jj}}.$$

Due to this plug-in estimate of σ^2, we need to rely on a special *Student's t distribution* to calculate the approximate 95% confidence interval, getting:

$$\hat{\beta}_j \pm t_{n-k, 0.975} \cdot SE(\hat{\beta}_j),$$

where $t_{n-k, 0.975}$ is the 97.5th percentile of the Student's t distribution with $n - k$ degrees of freedom.

Although, with large $n - k$ (e.g., $n - k \geq 30$), Student's t distribution becomes very similar to normal, so we can use the approximation

$$\hat{\beta}_j \pm z_{0.975} \cdot SE(\hat{\beta}_j) \quad \text{or} \quad \hat{\beta}_j \pm 1.96 \cdot SE(\hat{\beta}_j),$$

where $z_{0.975} \approx 1.96$ is the 97.5th percentile of the standard normal distribution.

When this confidence interval excludes 0, we can reject the statistical hypothesis that the corresponding coefficient is zero. This type of conclusion might, for example, lead to the inference that a drug has a statistically significant effect compared to a placebo in a clinical trial – or, conversely, that it does not differ from a placebo, potentially costing a pharmaceutical company a lot of money.

Personally, I prefer bootstrap confidence intervals for their simplicity! Fortunately, the above analytical formulas are implemented in most regression libraries, such as `statsmodels` , so we can use them out of the box. That said, I strongly recommend taking a dedicated course on regression analysis before developing confidence in interpreting regression results – the assumptions and nuances really matter here.

4.10 Historical note: Why "regression"?

The term *regression* was introduced by Francis Galton in the late 1800s while studying the relationship between the heights of parents and their children. He noticed that:

- Tall parents tended to have tall children, but not as tall as themselves.
- Short parents tended to have short children, but not as short as themselves.

In other words, children's heights tended to *regress* towards the mean of the general population. This phenomenon – the statistical tendency of extreme measurements to be followed by more average ones – was termed "regression toward mediocrity" by Galton.

From this, the statistical method used to quantify the relationship between variables – like parent and child height – became known as regression analysis.

Today, the term regression has broadened far beyond its original context. It now may refer to any method that models the relationship between a dependent (outcome / target) variable and one or more independent (input) variables, whether the outcome is continuous (as in linear regression) or categorical (as in logistic regression).

See [Wikh] for more discussion.

4.11 Discussion

In this chapter, we explored how to build simple predictive models for both continuous and categorical outcomes, and how to estimate their parameters from data. We also introduced the bootstrap method as a way to assess uncertainty in these parameter estimates. In addition, we examined how the mathematical structure of the underlying optimization problem influences our ability to efficiently and reliably find a globally optimal solution (e.g., in convex vs. non-convex optimization problems). As we will see next, neurons – key components of neural networks and large language models – echo the structure and logic of the simple regression models.

4.12 Further learning resources

- Regression models:
  - Statistical perspective:
    - * Frequentist [JW07; CB24].
    - * Bayesian [Gel+13].
    - * Non-parametric / Bayesian [EH21].
    - * Asymptotic (large-sample) [Vaa98].
  - Machine learning perspective: [BN06].
  - Causal inference perspective: [AP09], https://miguelhernan.org/whatifbook [HR20].
  - Panel / time series models: [Hay11; SSS00].
  - Discrete choice models: [Tra09].
  - *Simple Regression Analysis in Public Health* online course by John McGready on Coursera: https://www.coursera.org/learn/simple-regression-analysis-public-health.

- Convexity and gradient descent: [BV04].
- Bootstrap: [TE93; Vaa98].

- Extreme value theory: [Col+01; EKM13].
- Applied graph theory: [EK+10].
- General optimization: [KW19; NW06].
- Creating Python libraries: https://packaging.python.org/en/latest/tutorials/packaging-projects/.

4.13 References

[AP09]   Joshua D Angrist and Jörn-Steffen Pischke. *Mostly Harmless Econometrics: An Empiricist's Companion.* Princeton University Press, 2009.

[BF81]   Peter J Bickel and David A Freedman. "Some asymptotic theory for the bootstrap". In: *The Annals of Statistics* 9.6 (1981), pp. 1196–1217.

[BN06]   Christopher M Bishop and Nasser M Nasrabadi. *Pattern Recognition and Machine Learning.* Springer, 2006.

[BV04]   Stephen Boyd and Lieven Vandenberghe. *Convex Optimization.* Cambridge, UK: Cambridge University Press, 2004. URL: https://web.stanford.edu/~boyd/cvxbook/.

[CB24]   George Casella and Roger Berger. *Statistical Inference.* CRC Press, 2024.

[Chi+25]   Wayne Chi et al. "Copilot Arena: A platform for code LLM evaluation in the wild". In: *arXiv preprint arXiv:2502.09328* (2025).

[Chi+24]   Wei-Lin Chiang et al. "Chatbot Arena: An open platform for evaluating LLMs by human preference". In: *ICML.* 2024.

[Col+01]   Stuart Coles et al. *An Introduction to Statistical Modeling of Extreme Values.* Vol. 208. Springer, 2001.

[EK+10]   David Easley, Jon Kleinberg, et al. *Networks, Crowds, and Markets: Reasoning About a Highly Connected World.* Cambridge University Press, 2010.

[EHH96]   Bradley Efron, Elizabeth Halloran, and Susan Holmes. "Bootstrap confidence levels for phylogenetic trees". In: *PNAS* 93.23 (1996), pp. 13429–13429.

[EH21]   Bradley Efron and Trevor Hastie. *Computer Age Statistical Inference*. Cambridge University Press, 2021.

[EKM13]   Paul Embrechts, Claudia Klüppelberg, and Thomas Mikosch. *Modelling Extremal Events: For Insurance and Finance*. Springer Science & Business Media, 2013.

[For57]   Lester R Ford Jr. "Solution of a ranking problem from binary comparisons". In: *The American Mathematical Monthly* 64.8P2 (1957), pp. 28–33.

[Gel+13]   Andrew Gelman et al. *Bayesian Data Analysis*. Chapman and Hall/CRC, 2013.

[ggp]   ggplot2. *Diamonds*. URL: https://ggplot2.tidyverse.org/reference/diamonds.html (visited on 05/03/2025).

[Hay11]   Fumio Hayashi. *Econometrics*. Princeton University Press, 2011.

[HR20]   Miguel A Hernán and James M Robins. *Causal Inference: What If*. CRC Boca Raton, FL, 2020.

[Hun04]   David R Hunter. "MM algorithms for generalized Bradley-Terry models". In: *The Annals of Statistics* 32.1 (2004), pp. 384–406.

[JW07]   Richard A. Johnson and Dean W. Wichern. *Applied Multivariate Statistical Analysis*. 6th. Prentice Hall, 2007. Chap. 2.

[KW19]   Mykel J Kochenderfer and Tim A Wheeler. *Algorithms for Optimization*. MIT Press, 2019.

[Lan14]   Jared P Lander. *R for everyone: Advanced analytics and graphics*. Pearson Education, 2014.

[NW06]   Jorge Nocedal and Stephen J Wright. *Numerical Optimization*. Springer, 2006.

[SSS00]   Robert H Shumway, David S Stoffer, and David S Stoffer. *Time Series Analysis and Its Applications*. Springer, 2000.

[TE93]   Robert J Tibshirani and Bradley Efron. *An Introduction to the Bootstrap*. Chapman & Hall, 1993.

[Tra09]   Kenneth E. Train. *Discrete Choice Methods with Simulation*. 2nd. Cambridge, UK: Cambridge University Press, 2009. URL: https://eml.berkeley.edu/books/choice2.html.

[Vaa98]   A. W. van der Vaart. *Asymptotic Statistics*. Cambridge University Press, 1998.

[Wika]   Wikipedia. *Bradley-Terry model*. URL: https://en.wikipedia.org/wiki/Bradley%E2%80%93Terry_model (visited on 05/30/2025).

[Wikb]   Wikipedia. *Comma-separated values*. URL: https://en.wikipedia.org/wiki/Comma-separated_values (visited on 05/03/2025).

[Wikc]   Wikipedia. *Convex function*. URL: https://en.wikipedia.org/wiki/Convex_function (visited on 05/03/2025).

[Wikd]   Wikipedia. *Elo rating system*. URL: https://en.wikipedia.org/wiki/Elo_rating_system (visited on 05/30/2025).

[Wike]   Wikipedia. *Himmelblau's function*. URL: https://en.wikipedia.org/wiki/Himmelblau%27s_function (visited on 05/30/2025).

[Wikf]   Wikipedia. *Limited-memory BFGS*. URL: https://en.wikipedia.org/wiki/Limited-memory_BFGS (visited on 05/30/2025).

[Wikg]   Wikipedia. *Numerical differentiation*. URL: https://en.wikipedia.org/wiki/Numerical_differentiation (visited on 05/03/2025).

[Wikh]   Wikipedia. *Regression toward the mean*. URL: https://en.wikipedia.org/wiki/Regression_toward_the_mean (visited on 05/02/2025).

[Zhe+23]   Lianmin Zheng et al. "Judging LLM-as-a-judge with MT-Bench and Chatbot Arena". In: *NeurIPS* 36 (2023), pp. 46595–46623.

5 Deep learning

In this chapter, we review **deep learning** – a class of advanced prediction algorithms based on neural networks, which serve as the foundation for large language models. In particular, we focus on how neural networks can be trained using gradient descent.

5.1 A neuron: A linear model plus non-linearity

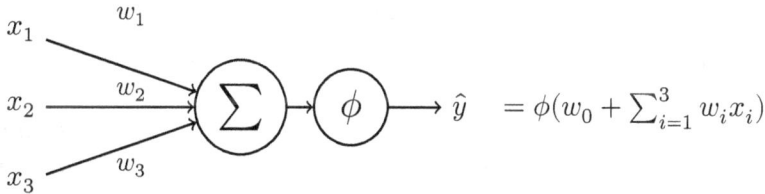

Figure 5.1: A neuron: a linear model with parameters w, inputs x, followed by a non-linear transformation $\phi(\cdot)$

A neuron is an equation that:

1. Takes input variables – e.g., x_1, x_2, x_3 for person's age, height, weight;
2. Multiplies them by parameters (numbers, weights) w_1, w_2, w_3;
3. Sums up these products and adds an intercept term w_0, which is also called *bias* in the deep learning world;
4. Passes the result through a non-linear *activation* function $\phi(\cdot)$:

$$\hat{y} = \phi\left(w_0 + \sum_{i=1}^{3} w_i x_i\right).$$

Steps 1 through 3 are effectively a linear regression equation.

Common activation functions:

- **Sigmoid (logistic):**

  - $\phi(x) = \sigma(x) = \frac{\exp(x)}{\exp(x)+1} = \frac{1}{1+\exp(-x)}$.
  - Effectively a binary logistic regression equation.
  - Maps any real number to the interval $(0, 1)$ and was historically popular. Today, it is more rarely used due to its relative slowness and *saturation* issues: when the output is near 0 or 1, the gradient becomes very small, which can stall learning during gradient descent.

- **ReLU (Rectified Linear Unit):**

  - $\phi(x) = \text{ReLU}(x) = \max(x, 0) = x \cdot \mathbf{1}(x > 0)$.
  - Takes a maximum between the input and a zero, setting any negative numbers to zero. This one is fast, works well in practice, and is very popular.
  - $\mathbf{1}(x > 0)$ is the elementwise indicator function. It returns 1s for positive elements of x and 0s otherwise.
    * *Note:* Technically, $\text{ReLU}(x)$ is not differentiable at $x = 0$, since the function has a kink / sharp corner at that point (the limit, which we use to define derivatives, does not exist there). However, in practice, we simply *define* the derivative at that point as $\frac{d\text{ReLU}}{dx}(0) = 0$, and this works perfectly well in training neural networks. For the curious, this choice can be theoretically justified using the more advanced concept of *sub-gradients*, which generalize derivatives to non-differentiable convex functions [Wikc]. (The ReLU activation function itself is convex.)

- **No activation (identity):**

  - $\phi(x) = x$.
  - Neurons without an activation transformation are sometimes used at the output layer of the neural net (more on this later).

- Many other activation functions, such as *leaky ReLU*, *tanh*, and *GELU*, can also be used – see [Wikb] for more options.

Initially, a neuron's weights w_i are unknown – and are randomly initialized, to be fine-tuned later.

Using ReLU, the neuron in our example computes:

$$f(x) = \max(w_0 + \sum_{i=1}^{3} w_i x_i, 0).$$

In code:

```python
import numpy as np

# example - a single input observation (3 variables)
x = np.array([-5.2, -0.7, 0.3])

# initializing parameters for a single neuron via standard
  ↪ normal distribution
np.random.seed(42)
w = np.random.randn(3)   # 3 weights
b = np.random.randn()    # bias (intercept)

# ReLU
def relu(x):
    return np.maximum(x, 0)

# computation
z = np.dot(w, x) + b
out = relu(z)

print("Weights w:", w)
print("Bias b:", b)
print("Linear output (z):", z)
print("ReLU activation:", out)
```

```
Weights w: [ 0.49671415 -0.1382643   0.64768854]
Bias b: 1.5230298564080254
Linear output (z): -0.7687921670003479
ReLU activation: 0.0
```

In the neural net literature, typically, the *bias / intercept term* is treated as a conceptually separate entity from the inputs to the neuron that are multiplied by the neuron's weights; intercept is considered to be *implicit* in the neuron and is usually not explicitly indicated on the computational graph diagrams, such as Figure 5.1 and Figure 5.2.

5.2 Neural networks

Hidden layer

Input layer   **Output layer**

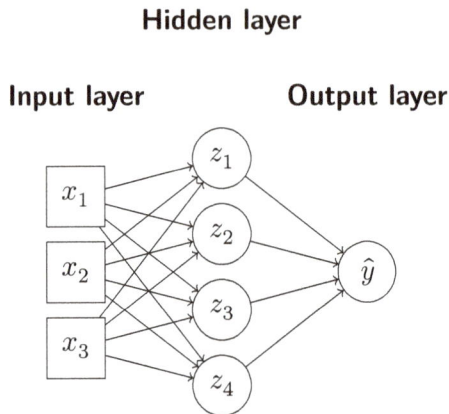

Figure 5.2: A simple neural net

A **neural network (neural net)** is a stack of neurons organized in layers. In the diagram above, the square nodes represent the raw data input for the neural net, while the round nodes represent the neurons. Each arrow represents a neuron's weight multiplying one of its inputs.

- **Input layer:** The first vertical layer on the left consists of 3 square nodes representing 3 raw data inputs x_1, x_2, x_3 (e.g., age, weight, height variables).
- **Hidden layer:** The second vertical layer consists of 4 round nodes that represent 4 neurons z_1 to z_4, each computing a ReLU-transformed linear combination of inputs, using its own weights. As this layer is on the inside of the net, it can sometimes be called a **hidden** layer.
- **Output layer:** The final layer of a neural net, on the right, consists of a single round node (neuron) computing prediction $\hat{y}$ from hidden layer outputs.

Every neuron – round node – has its own parameters multiplying the inputs to the neuron, which are indicated by the arrows. For example:

- $z_1 = \max(w_{01}^{(1)} + w_{11}^{(1)} x_1 + w_{21}^{(1)} x_2 + w_{31}^{(1)} x_3, 0)$.
- $z_2 = \max(w_{02}^{(1)} + w_{12}^{(1)} x_1 + w_{22}^{(1)} x_2 + w_{32}^{(1)} x_3, 0)$.
- And so on.

Each neuron in the hidden layer in this example has 4 own parameters – 3 for input variables and one for the intercept / bias. Because there are 4 neurons in the hidden layer, the hidden layer produces 4 outputs – one for each neuron – and, altogether, the hidden layer has 16 tweakable parameters (4 for each neuron in the layer).

> 💡 Tip
>
> A **fully connected layer** means every neuron in a layer connects to all outputs from the previous layer.

The final layer – called the output layer – contains one neuron $\hat{y}$ in our example (although a neural net can as well have multiple neurons in the output, generating multiple predictions simultaneously). The output neuron takes as input 4 outputs from the preceding (hidden) layer and multiplies them by its own parameters.

Neurons in the output layer, in contrast to hidden layer neurons, may be pure linear transformations (without the non-linearity) – or they may have some distinct transformation happening to their outputs, like sigmoid or softmax. In our example, we will assume a pure linear model at the output:

- $\hat{y} = w_{01}^{(2)} + w_{11}^{(2)} z_1 + w_{21}^{(2)} z_2 + w_{31}^{(2)} z_3 + w_{41}^{(2)} z_4$.

We can see the output neuron has 5 parameters – 4 for inputs and 1 for the bias – and no non-linear transformation at the output.

For example, if the presented neural net takes age, weight, and height of a person as input, at the output it could be trying to predict some characteristic about that person's health.

At the beginning, we do not know what the values of the parameters of all the neurons should be to make an accurate prediction – so we can just randomly initialize them at the start – and then figure how to fine-tune them later on – to yield the best prediction possible.

5.2.1 Matrix formulation of forward computation

While seeing individual parameters of each neuron is helpful at the early stage to understand what is happening in the neural net, it quickly becomes unwieldy – especially if the number and the size of the hidden layers grows. It turns out linear algebra comes to save the day again – all the computation we have been doing forward from the inputs towards the outputs can be represented as matrix dot products with non-linear transformations mixed in!

In our example, we have **two layers** performing the computation and we will define 2 matrices for each computational layer – to hold their parameters. Here are the key matrices:

- Input:

    - A single observation ($n = 1$): x is 1×3 vector containing 3 input variables the neural net operates on – e.g., age, weight, height for a single person.

- Layer 1 | Input dimension: 3 (data points); Output dimension: 4 (neurons)

    - W_1 is 3×4 parameter matrix. Each column in it represents 3 parameters of one of the neurons in the first layer – that the neuron uses to multiply inputs.
    - b_1 is 1×4 parameter matrix containing bias / intercept parameters of each neuron.

- Layer 2 | Input dimension: 4 (preceding layer neurons); Output dimension: 1 (outputs)

    - W_2 is 4×1 parameter matrix. Each column in it represents 4 parameters of one of the output neurons – that the neuron uses to multiply 4 outputs from the preceding layer.
    - b_2 is 1×1 parameter matrix containing bias / intercept parameter of the single output neuron.

Having these matrices set up, we can now write down the neural net operation in matrix form:

$$Z = \text{ReLu}(x \cdot W_1 + b_1) = \max(x \cdot W_1 + b_1, 0)$$
$$\hat{y} = Z \cdot W_2 + b_2$$

As a reminder, here $\cdot$ means a matrix dot product, $+$ is elementwise addition, $\max(\cdot, 0)$ sets every negative element of the input array to zero.

Or we can also write this as a one-line function $N(x)$, specifying dimensions of matrices throughout dot product and other computations:

$$\hat{y} = N(x) = \underbrace{\max(\underbrace{x}_{1\times3} \cdot \underbrace{W_1}_{3\times4} + \underbrace{b_1}_{1\times4}, 0)}_{\underbrace{}_{1\times4}} \cdot \underbrace{W_2}_{4\times1} + \underbrace{b_2}_{1\times1}.$$

So, $N(x)$ takes as input 1×3 observation and outputs 1×1 prediction.

And this is all there is to a neural net forward computation – a sequence of linear algebra operations with non-linearities in-between!

From 1 to n observations

We can extend the neural network to handle multiple observations at once by stacking them as rows in a matrix X of shape $n \times 3$, where n is the number of observations and each row contains 3 input variables.

Then the function $N(X)$ becomes:

$$\hat{y} = N(X) = \underbrace{\max(\underbrace{X}_{n\times3} \cdot \underbrace{W_1}_{3\times4} + \underbrace{b_1}_{1\times4}, 0)}_{\underbrace{}_{n\times4}} \cdot \underbrace{W_2}_{4\times1} + \underbrace{b_2}_{1\times1}.$$

Here is where *broadcasting* comes in. For example, the plus $+$ operator between an $n \times 4$ matrix $X \cdot W_1$ and a 1×4 bias vector b_1 automatically adds the bias vector to each row of the matrix. This behavior – adding a smaller array to a larger one along a matching dimension – is called broadcasting, and it is supported in Python libraries like NumPy and PyTorch (we will encounter PyTorch later).

As a result, the neural net function $N(X)$ takes as input an $n \times 3$ observation matrix X and outputs an $n \times 1$ prediction matrix – containing one prediction for each of n input observations.

In code:

```python
import numpy as np

# example - n=5 observations, each with 3 input variables
X = np.array([
    [1.2, -0.7, 0.3],
    [0.5,  1.1, -1.4],
    [-0.3, 0.8, 0.9],
    [2.0, -1.0, 0.1],
    [0.0, 0.0, 0.0]
])  # shape: (5, 3)

# random weights and biases for two layers of neurons
np.random.seed(42)
W1 = np.random.randn(3, 4)     # shape: (3, 4)
b1 = np.random.randn(1, 4)     # shape: (1, 4)
W2 = np.random.randn(4, 1)     # shape: (4, 1)
b2 = np.random.randn(1, 1)     # shape: (1, 1)

# activation function
def relu(x):
    return np.maximum(0, x)

# forward pass / computation
Z = relu(X @ W1 + b1)          # shape: (5, 4)
y_hat = Z @ W2 + b2            # shape: (5, 1)

# X @ W1 is a dot product, same as np.dot(X, W1)

print("Predictions (y_hat):", y_hat)
```

```
Predictions (y_hat): [[-0.2375191 ]
 [-2.72450733]
 [ 1.46564877]
 [-2.33341581]
 [ 1.22058185]]
```

5.2.2 Why neural nets?

While a linear regression is a powerful and simple predictive tool, it has a fundamental limitation – it can only learn linear relationships between variables. When data exhibits more complex, non-linear patterns, a straight line through the cloud of points often won't cut it.

This limitation can be somewhat addressed via "feature engineering," where we include as model input not only raw variables, but also their non-linear transformations, such as squared variables or products among variables. The output variable can also be transformed, for example, by taking its logarithm. This can allow a linear regression to model non-linear dependencies between the raw input variables and the raw output variable – while remaining a linear model because it is *linear in the estimated coefficients*. However, manually hand-crafting features to include in the model is a trial-and-error process – it does not scale well as the number of variables grows and can be very time-consuming.

Neural networks offer a more flexible alternative – they learn complex features automatically through layers of computation. A crucial component that gives neural nets their flexibility is the **non-linear activation function** applied within each hidden layer neuron. Without this, the network's output would be just a sequence of linear transformations, collapsing back into a single linear model – no more expressive than ordinary linear regression. It is the non-linearity that enables neural networks to approximate complex, curved relationships in data.

Figure 5.3 illustrates this idea. The data set clearly follows a non-linear pattern that a linear regression line (in red), based on a raw untransformed input variable, fails to capture, leaving large prediction errors. A neural net (dashed blue curve), on the other hand, fits the pattern nearly perfectly.

A key theoretical result called the **Universal Approximation Theorem** shows that a neural network with just a single hidden layer, involving an

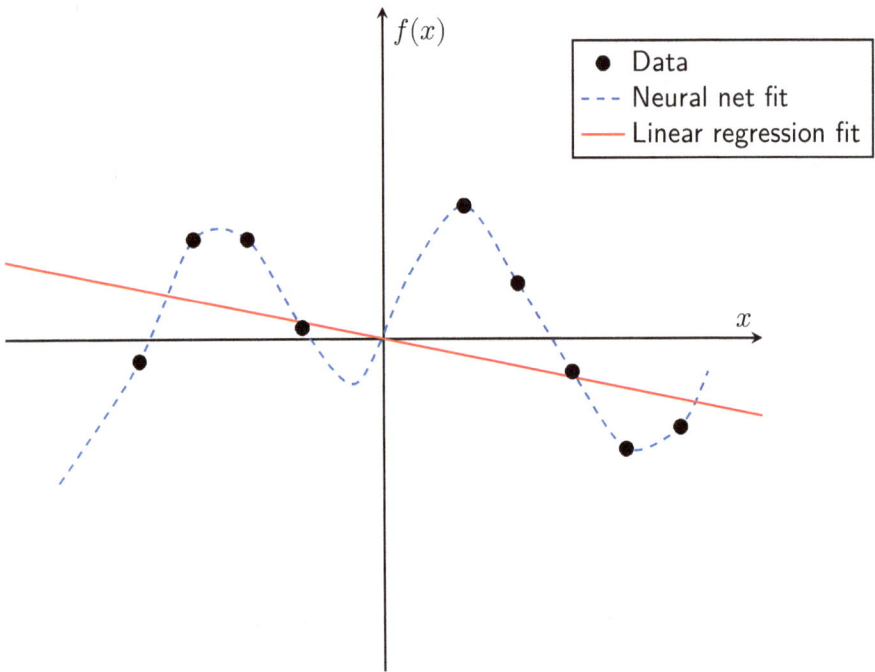

Figure 5.3: Fitting a pattern: Linear regression vs. neural net

appropriate non-linear activation function, such as ReLU, can approximate any continuous function to arbitrary accuracy – provided the layer has enough neurons. See [Wikd] for more details.

This result is significant – it suggests that neural nets, in principle, can learn a broad class of systematic patterns – such as, perhaps, how words tend to follow one another – if given sufficient data and capacity and as long as the patterns are expressible as continuous functions on suitably encoded inputs.

In practice, *deep* neural networks (those with *multiple* hidden layers) tend to work better than shallow ones. Depth – not just width – enables a model to build up complex representations of data through layer-by-layer transformations. As a rule of thumb, the deeper the net, the more powerful it becomes – at least, up to a point.

---

i Deep vs. shallow

Technically, any neural network with more than one hidden layer qualifies as "deep." By that definition, our running example is a shallow net – it has only a single hidden layer. In practice, though, most practitioners reserve the term deep for networks with many layers and millions or billions of parameters, like those used in large-scale models and language systems.

5.3 Training neural nets

Let us go back to our example neural network (Figure 5.4). For some starting values of W_1, b_1, W_2, and b_2, for $n \times 3$ input data X (i.e., n observations, each with 3 input variables), our predicted output is $n \times 1$ matrix:

$$\hat{y} = \max(X \cdot W_1 + b_1, 0) \cdot W_2 + b_2.$$

As in linear regression, we can compute the mean squared error (MSE) between the prediction and true data:

$$\mathrm{MSE}(W_1, b_1, W_2, b_2) = (y - \hat{y})^T (y - \hat{y})/n.$$

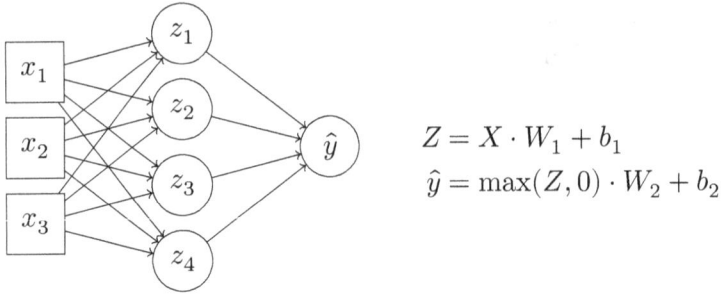

Figure 5.4: A simple neural net template

We now want to fine-tune all the neural net parameters to minimize this MSE objective. Unfortunately for us, due to the complexity of the neural net function, there is no closed-form analytical solution. Nevertheless, we already know, from the previous chapter, a way to search for the parameters that minimize this MSE – compute partial derivatives of MSE with respect to parameters of the neural net and apply *gradient descent*. If we flatten all parameters in the neural net into a single parameter vector, we get:

$$\text{parameters}_{t+1} = \text{parameters}_t - \gamma \nabla_{\text{parameters}} \text{MSE}.$$

Here, γ is the learning rate, and $\nabla_{\text{parameters}}\text{MSE}$ represents the gradient vector of the loss with respect to all parameters.

Notice that, due to ReLU non-linearity, the neural net MSE function is no longer quadratic or convex with respect to the neural net's parameter vector (unlike the MSE in linear regression). Therefore, while the gradient descent iteration might arrive at some stationary point, such as an MSE local minimum, there is certainly *no theoretical guarantee* the resulting parameter combination would be a *global* minimum of MSE. Thus, we should always keep in mind when working with neural nets that there might exist a parameter vector better than the one found via gradient descent – we just cannot be sure.

Also note that parameter *identification* (optimal solution uniqueness) is generally not achievable in the context of neural networks – neural nets are *overparametrized* – many different parameter sets can produce the same loss value. One way to demonstrate this is via neuron permutation: exchanging two neurons within a hidden layer – along with their respective input and output connections – leaves the network's output unchanged and yields the same loss, yet produces a distinct (reshuffled) parameter array. As a result, small changes in the training data can lead to very different estimates for a specific parameter.

This non-identifiability and potential differences in the estimated parameter sets across different training runs mean, for instance, that the bootstrap estimates of the neural net's parameter confidence intervals would not be very meaningful. Plus, in general, with models potentially having billions of parameters, it is not practical to analyze uncertainty for each parameter. That is why deep learning papers rarely report confidence intervals for individual parameters. (Bootstrap confidence intervals for the model's error / loss are still meaningful though – and we will look at those later.) That said, methods to get some idea about the neural net parameter uncertainty do exist. For example, the *stochastic weight averaging-Gaussian* approach [Mad+19] estimates parameter uncertainty by modeling how the weights vary across different snapshots collected during the final stages of training.

Acknowledging these issues, let us proceed with the gradient descent method. To efficiently compute the partial derivatives for all layer parameters in the neural net, we will rely on a method called *backpropagation*. This technique applies the *chain rule of calculus* (which dates back at least to Lagrange – 18th century) to systematically work backwards through the network – from the loss function, through layers, to each layer's parameters – computing derivatives at each step.

5.3.1 Computing derivatives manually

We will start by deriving every gradient of our toy two-layer neural network by hand. First we handle a single $n = 1$ training example; afterwards we extend the result to a batch of $n \geq 1$ examples. In practice, deep-learning libraries, such as PyTorch, compute gradients automatically via automatic differentiation [Wika], so you can, and usually should, skip the algebra. Nevertheless, I will demonstrate the procedure so that you can appreciate what happens under the hood of these libraries.

The equation for our neural net is:

$$\hat{y} = \max(X \cdot W_1 + b_1, 0) \cdot W_2 + b_2.$$

We start with the case of $n = 1$ single-observation input, which allows us not to worry about some issues with broadcasting, at least, for now. We break the computation into elementary pieces:

$$A = X \cdot W_1 + b_1$$
$$Z = \max(A, 0) = A \odot \mathbf{1}(A > 0)$$
$$\hat{y} = Z \cdot W_2 + b_2$$
$$\epsilon = y - \hat{y}$$
$$\text{MSE} = \epsilon^2/n \underset{n=1}{=} \epsilon^2.$$

Here $\mathbf{1}(A > 0)$ has the same shape as A and is a 1 / 0 elementwise indicator for whether elements of array A are positive, so $\max(A, 0) = A \odot \mathbf{1}(A > 0)$, where $\odot$ is an elementwise product of two arrays. Letter ϵ is pronounced as *epsilon*.

We will now compute derivatives for each of the steps, going in reverse:

$$\text{MSE} = \epsilon^2 \quad \rightarrow \quad \frac{\partial \text{MSE}}{\partial \epsilon} = 2\epsilon = 2(y - \hat{y})$$

$$\epsilon = y - \hat{y} \quad \rightarrow \quad \frac{\partial \epsilon}{\partial \hat{y}} = -1$$

$$\hat{y} = Z \cdot W_2 + b_2 = \begin{bmatrix} z_1 & z_2 & z_3 & z_4 \end{bmatrix}_{Z:(1\times4)} \cdot \begin{bmatrix} w_{11}^{(2)} \\ w_{21}^{(2)} \\ w_{31}^{(2)} \\ w_{41}^{(2)} \end{bmatrix}_{W_2:(4\times1)} + \begin{bmatrix} w_{01}^{(2)} \end{bmatrix}_{b_2:(1\times1)}$$

$$\rightarrow \quad \frac{\partial \hat{y}}{\partial w_{i1}^{(2)}} = z_i \quad \forall i \in 1:4;$$

$$\rightarrow \quad \frac{\partial \hat{y}}{\partial w_{01}^{(2)}} = 1;$$

$$\rightarrow \quad \frac{\partial \hat{y}}{\partial z_i} = w_{i1}^{(2)} \quad \forall i \in 1:4,$$

where $\forall$ symbol means *for all*.

For compactness, we could put these partial derivatives into vectors to get gradients:

$$\rightarrow \quad \nabla_{W_2}\hat{y} = \begin{bmatrix} \frac{\partial \hat{y}}{\partial w^{(2)}_{11}} \\ \frac{\partial \hat{y}}{\partial w^{(2)}_{21}} \\ \frac{\partial \hat{y}}{\partial w^{(2)}_{31}} \\ \frac{\partial \hat{y}}{\partial w^{(2)}_{41}} \end{bmatrix} = \begin{bmatrix} z_1 \\ z_2 \\ z_3 \\ z_4 \end{bmatrix} ;$$

$$\rightarrow \quad \nabla_{b_2}\hat{y} = \begin{bmatrix} \frac{\partial \hat{y}}{\partial w^{(2)}_{01}} \end{bmatrix} = \begin{bmatrix} 1 \end{bmatrix} ;$$

$$\rightarrow \quad \nabla_{z}\hat{y} = \begin{bmatrix} \frac{\partial \hat{y}}{\partial z_1} & \frac{\partial \hat{y}}{\partial z_2} & \frac{\partial \hat{y}}{\partial z_3} & \frac{\partial \hat{y}}{\partial z_4} \end{bmatrix} = \begin{bmatrix} w^{(2)}_{11} & w^{(2)}_{21} & w^{(2)}_{31} & w^{(2)}_{41} \end{bmatrix}.$$

For example, $\nabla_{W_2}\hat{y}$ tells us how prediction $\hat{y}$ (which is scalar, given that, for now, $n = 1$) changes with a change in parameters W_2. The shape of a gradient of a scalar function with respect to some variable should have the same shape as the variable itself – e.g., $\nabla_{W_2}\hat{y}$ should be the same shape as W_2, which helps check for correctness the shapes of gradient arrays.

Now we want to compute the derivatives of MSE with respect to W_2 parameters. Here, to our salvation comes the **chain rule of calculus**.

💡 Chain rule for single-variable functions

If we have a function inside a function $f(x) = g(h(x))$, then the chain rule tells us how to take the derivative of the whole thing: $\frac{df}{dx} = \frac{dg}{dh} \cdot \frac{dh}{dx}$. For example, let $h(x) = 2x$ and $g(h) = h^2$. Then $f(x) = g(h(x)) = (2x)^2 = 4x^2$. Now, we apply the chain rule: $\frac{dh}{dx} = 2$, $\frac{dg}{dh} = 2h = 2 \cdot (2x) = 4x$. So $\frac{df}{dx} = \frac{dg}{dh} \cdot \frac{dh}{dx} = 4x \cdot 2 = 8x$.

In our case, we know:

- $\frac{\partial \text{MSE}}{\partial \epsilon} = 2(y - \hat{y})$
- $\frac{\partial \epsilon}{\partial \hat{y}} = -1$
- $\frac{\partial \hat{y}}{\partial w^{(2)}_{i1}} = z_i$

Then, by chain rule (adapted for the partial derivative case),

$$\frac{\partial \text{MSE}}{\partial w^{(2)}_{i1}} = \frac{\partial \text{MSE}}{\partial \epsilon} \frac{\partial \epsilon}{\partial \hat{y}} \frac{\partial \hat{y}}{\partial w^{(2)}_{i1}} = -2z_i(y - \hat{y}) = 2z_i(\hat{y} - y).$$

By implication, the gradient of MSE with respect to W_2 is thus

$$\nabla_{W_2}\text{MSE} = \begin{bmatrix} 2z_1(\hat{y} - y) \\ 2z_2(\hat{y} - y) \\ 2z_3(\hat{y} - y) \\ 2z_4(\hat{y} - y) \end{bmatrix}_{4 \times 1}.$$

This is brilliant! Knowing intermediate derivatives for a sequence of computations, we can figure out the derivative of the output value with respect to any variable entered into the computation! The above equation is exactly what we need to run our gradient update step on W_2 part of the parameter set of the neural net. For instance, it tells us that if we increment second scalar parameter in W_2 by 1, MSE can be expected to increase by $2z_2(\hat{y} - y)$ amount (which is just a number we can compute – all its constituent parts are known after forward computation on the neural net has been completed).

By analogy, from:

- $\frac{\partial \hat{y}}{\partial w_{01}^{(2)}} = 1;$
- $\frac{\partial \hat{y}}{\partial z_i} = w_{i1}^{(2)} \quad \forall i \in 1:4;$

we similarly get the gradient of MSE with respect to the intercept parameter in the second layer:

$$\nabla_{b_2}\text{MSE} = \begin{bmatrix} 2(\hat{y} - y) \end{bmatrix}_{1 \times 1}.$$

The gradient of MSE with respect to the four outputs Z of the first layer for a single observation is:

$$\nabla_Z\text{MSE} = \begin{bmatrix} 2w_{11}^{(2)}(\hat{y} - y) & 2w_{21}^{(2)}(\hat{y} - y) & 2w_{31}^{(2)}(\hat{y} - y) & 2w_{41}^{(2)}(\hat{y} - y) \end{bmatrix}_{1 \times 4}.$$

We now know all the partial derivatives of MSE with respect to the parameters W_2 and b_2 of the second layer as well as derivatives of MSE with respect to the outputs of the first layer Z.

All that is left is to compute derivatives of MSE with respect to parameters of the first layer. Thanks to the chain rule, because we already know $\nabla_Z\text{MSE}$, all that is left is to compute $\nabla_{W_1}Z$ and $\nabla_{b_1}Z$, from which we can compute $\nabla_{W_1}\text{MSE} = \nabla_{W_1}Z \cdot \nabla_Z\text{MSE}$ and $\nabla_{b_1}\text{MSE} = \nabla_{b_1}Z \cdot \nabla_Z\text{MSE}$.

Here Z is a multi-output function, so, technically, for example, $\nabla_{W_1} Z$ would represent an array of partial derivatives of every output value with respect to every input variable. Unfortunately, notation for this varies a lot across authors. Some authors would call such an array of partial derivatives a *Jacobian*, not a gradient – reserving the term gradient for scalar-valued functions, while using the term *Jacobian* for vector-valued outputs. As a further complication, the Jacobian is formally defined as a 2D matrix, so indexing it properly would require flattening the input and output array shapes to vectors. But doing so would obscure the original structure of the arrays (e.g., W_1) and complicate notation.

To avoid this confusion and keep the connection with code more intuitive, I will refer to any array of partial derivatives as a "gradient," and continue using the nabla ∇ notation. I will also preserve input and output shapes when forming the partial derivative arrays, meaning the resulting "gradients" may be higher-dimensional tensors. This simplifies formulas, avoids unnecessary reshaping, and aligns directly with how most autodiff libraries (e.g., PyTorch) behave in practice.

The presence of multiple outputs in Z also means we need to add up influence of parameters on MSE going through different Z neurons, which involves a dot product. There is a multivariate version of the chain rule that spells this out explicitly, but we won't discuss it here in detail, as the intuition should be clear as is, and the rule is complicated enough to get confusing. Unfortunately, multivariate calculus gets a bit tedious in its detail.

Recalling the first layer forward computation:

$$A = X \cdot W_1 + b_1$$

$$= \begin{bmatrix} x_1 & x_2 & x_3 \end{bmatrix}_{X:(1\times 3)} \cdot \begin{bmatrix} w_{11}^{(1)} & w_{12}^{(1)} & w_{13}^{(1)} & w_{14}^{(1)} \\ w_{21}^{(1)} & w_{22}^{(1)} & w_{23}^{(1)} & w_{24}^{(1)} \\ w_{31}^{(1)} & w_{32}^{(1)} & w_{33}^{(1)} & w_{34}^{(1)} \end{bmatrix}_{W_1:(3\times 4)}$$

$$+ \begin{bmatrix} w_{01}^{(1)} & w_{02}^{(1)} & w_{03}^{(1)} & w_{04}^{(1)} \end{bmatrix}_{b_1:(1\times 4)}$$

$$Z = \max(A, 0) = A \odot \mathbf{1}(A > 0),$$

we get, because of how the partial derivatives distribute across dot products:

- $\nabla_A Z = \mathbf{1}(A > 0)_{1\times 4} = \mathbf{1}(Z > 0)_{1\times 4}$;

- $\nabla_{W_1} A = \begin{bmatrix} X.T & X.T & X.T & X.T \end{bmatrix}_{3\times 4} = \begin{bmatrix} x_1 & x_1 & x_1 & x_1 \\ x_2 & x_2 & x_2 & x_2 \\ x_3 & x_3 & x_3 & x_3 \end{bmatrix}_{3\times 4}$;

- $\nabla_{b_1} A = \begin{bmatrix} 1 & 1 & 1 & 1 \end{bmatrix}_{1\times 4}$.

And so, finally, we get:

$$\nabla_{W_1} \text{MSE} = \nabla_{W_1} A \odot \nabla_A Z \odot \nabla_Z \text{MSE}$$

$$= \begin{bmatrix} x_1 & x_1 & x_1 & x_1 \\ x_2 & x_2 & x_2 & x_2 \\ x_3 & x_3 & x_3 & x_3 \end{bmatrix}_{3\times 4} \odot$$

$$\odot \begin{bmatrix} 2w_{11}^{(2)}(\hat{y}-y)(z_1>0) & 2w_{21}^{(2)}(\hat{y}-y)(z_2>0) & 2w_{31}^{(2)}(\hat{y}-y)(z_3>0) & 2w_{41}^{(2)}(\hat{y}-y)(z_4>0) \end{bmatrix}_{1\times 4}$$

$$= \begin{bmatrix} x_1 2w_{11}^{(2)}(\hat{y}-y)(z_1>0) & x_1 2w_{21}^{(2)}(\hat{y}-y)(z_2>0) & x_1 2w_{31}^{(2)}(\hat{y}-y)(z_3>0) & x_1 2w_{41}^{(2)}(\hat{y}-y)(z_4>0) \\ x_2 2w_{11}^{(2)}(\hat{y}-y)(z_1>0) & x_2 2w_{21}^{(2)}(\hat{y}-y)(z_2>0) & x_2 2w_{31}^{(2)}(\hat{y}-y)(z_3>0) & x_2 2w_{41}^{(2)}(\hat{y}-y)(z_4>0) \\ x_3 2w_{11}^{(2)}(\hat{y}-y)(z_1>0) & x_3 2w_{21}^{(2)}(\hat{y}-y)(z_2>0) & x_3 2w_{31}^{(2)}(\hat{y}-y)(z_3>0) & x_3 2w_{41}^{(2)}(\hat{y}-y)(z_4>0) \end{bmatrix}_{3\times 4}.$$

Using dot product, we can also more elegantly re-write

$$\nabla_{W_1} \text{MSE} = X_{3\times 1}^T \cdot (\nabla_A Z \odot \nabla_Z MSE)_{1\times 4},$$

which removes the need to manually create stacked / repeated matrix $\nabla_{W_1} A$.

Lastly:

$$\nabla_{b_1} \text{MSE} = \nabla_{b_1} A \odot \nabla_A Z \odot \nabla_Z \text{MSE}$$

$$= \begin{bmatrix} 2w_{11}^{(2)}(\hat{y}-y)(z_1>0) & 2w_{21}^{(2)}(\hat{y}-y)(z_2>0) & 2w_{31}^{(2)}(\hat{y}-y)(z_3>0) & 2w_{41}^{(2)}(\hat{y}-y)(z_4>0) \end{bmatrix}_{1\times 4}.$$

As you can see, this is incredibly gory (and very error-prone) to do manually – even for the simplest neural net. And we have only done it for a prediction from a single observation (row)!

Luckily, MSE is just the average of errors across observations – so we could sequentially compute gradients of MSE with respect to parameters for each observation using our equations and take the average gradient across observations for each parameter. So, all in all, these gradient expressions are sufficient to do gradient descent for our neural net.

Even more luckily, in practice, we would never have to do these computations by hand – software will handle these in a black-box manner for us – all we have to do is define a forward computation.

5.3.2 Derivatives for n observations

In practice, it can be faster to use matrix calculations directly on n observations, so, for completeness, I derive these matrix gradient expressions as well (without going to the level of scalar expressions though). Feel free to skip the equation part of this section if it puts too much fear into your heart.

Forward computation:

$$A = X \cdot W_1 + b_1$$
$$Z = \max(A, 0) = A \odot \mathbf{1}(A > 0)$$
$$\hat{y} = Z \cdot W_2 + b_2$$
$$\epsilon = y - \hat{y}$$
$$\text{MSE} = \epsilon^T \epsilon / n.$$

Backward computation (backpropagation):

$$\nabla_\epsilon \text{MSE} = \frac{2}{n}\epsilon = \frac{2}{n}(y - \hat{y}) \quad (n \times 1)$$

$$\nabla_{\hat{y}} \text{MSE} = \nabla_{\hat{y}}\epsilon \cdot \nabla_\epsilon \text{MSE} = -\nabla_\epsilon \text{MSE} = -\frac{2}{n}(y - \hat{y}) = \frac{2}{n}(\hat{y} - y) \quad (n \times 1)$$

$$\nabla_{\hat{y}}\epsilon = -I \quad (n \times n) \text{ negative identity matrix}$$

$$\nabla_{W_2} \text{MSE} = \nabla_{W_2}\hat{y} \cdot \nabla_{\hat{y}}\text{MSE} = \frac{2}{n}Z^T(\hat{y} - y) \quad (4 \times 1)$$

$$\nabla_{b_2} \text{MSE} = \nabla_{b_2}\hat{y} \cdot \nabla_{\hat{y}}\text{MSE} = \frac{2}{n}\mathbf{1}_{1 \times n} \cdot (\hat{y} - y) \quad (1 \times 1)$$

$$\mathbf{1}_{1 \times n} \text{ is a matrix of ones } (1 \times n)$$

$$\nabla_Z \text{MSE} = \nabla_Z \hat{y} \cdot \nabla_{\hat{y}} \text{MSE} = \frac{2}{n}(\hat{y} - y) \cdot W_2^T \quad (n \times 4)$$

$$\nabla_{W_1} \text{MSE} = \nabla_{W_1} A \cdot \nabla_A Z \cdot \nabla_Z \text{MSE}$$

$$= X^T \cdot ((Z > 0) \odot \nabla_Z \text{MSE})$$

$$= \frac{2}{n} X^T \cdot ((Z > 0) \odot (\hat{y} - y) \cdot W_2^T) \quad (3 \times 4)$$

$$\nabla_{b_1} \text{MSE} = \nabla_{b_1} A \cdot \nabla_A Z \cdot \nabla_Z \text{MSE}$$

$$= \mathbf{1}_{1 \times n} \cdot ((Z > 0) \odot \nabla_Z \text{MSE})$$

$$= \frac{2}{n} \mathbf{1}_{1 \times n} \cdot ((Z > 0) \odot (\hat{y} - y) \cdot W_2^T) \quad (1 \times 4).$$

What a fun series of equations!

Take-aways:

- Even a network this small produces such tedious algebra. Doing such computations by hand is painful and error-prone. We are lucky that modern software libraries compute all these derivatives automatically and all we have to do is define a forward computation.
- The shape of a gradient array for a scalar-valued function always matches the shape of the variable array the gradient is taken with respect to; this is a valuable sanity check when you implement new architectures.

---

i Dot products of multidimensional arrays

Unlike 2D matrices – where dot products are unambiguous – the exact way dot products between multidimensional arrays are computed depends on the conventions of specific libraries like PyTorch or NumPy. When translating the above equations to code, you may need to adjust some array shapes using reshaping or transposing to make sure they align correctly. You will see the exact, unambiguous code implementation later in the chapter.

5.4 Overfitting, regularization, hold-out evaluation

We have noted earlier the expressive power of neural nets compared to linear regression. With it comes a problem – a potential for **overfitting**.

When a model overfits, it performs well on the training data but poorly on new, previously unseen data. It has effectively memorized the training set – including its noise, outliers, and idiosyncrasies – rather than learning generalizable patterns. Re-examine Figure 5.3 – do you see a potential problem? Because neural nets can represent highly complex functions, they are especially prone to overfitting compared to simpler models like linear regression.

To mitigate overfitting, several techniques are commonly used:

- **Train / test split (or hold-out evaluation):** Always evaluate model performance on data it has not seen during training. This is typically done by splitting your data set into training and test sets (e.g., 80% / 20%). During training, we compute gradients and update parameters only on the training set, while we evaluate prediction error (e.g., MSE) on both training and test sets. For extra defense against overfitting, a separate **validation** set (a third split) can be used to tune model hyperparameters, while the test set is held back for final performance reporting. Another strategy is **cross-validation**, where the data is split multiple times into training and validation sets to average over different train / test partitions.

- **Regularization:** Regularization means augmenting the loss function to optimize the neural net parameters not just for the focal objective such as MSE prediction error, but also for the auxiliary objective, such as parameter magnitude – both of which are meant to be minimized. Let w designate a vector containing all (flattened) parameters of a neural net. Then our augmented loss function to take derivatives of could become $MSE + \lambda w^T w$, where $w^T w$ is just the sum of squared neural net parameter values and $\lambda > 0$ (pronounced as *lambda*) is a regularization strength parameter we set. This is known as L2, l_2, or *ridge* regularization. Optimizing for this loss function would mean simultaneously trying to reduce the prediction error and to keep the magnitude of parameters w relatively small, close to zero. Such objective augmentation can help avoid overfitting and can generally make training process more stable. (Usually, regularization is not applied to intercept parameters.)

In the context of our example neural net, regularization would be applied to W_1 and W_2, so overall loss could be written as $\mathcal{L} = MSE + \lambda(\text{sum}(W_1 \odot W_1) + \text{sum}(W_2 \odot W_2))$ and then the gradients of the loss with respect to parameters are just $\nabla_{W_1} MSE + 2\lambda W_1$ and $\nabla_{W_2} MSE + 2\lambda W_2$. The factor of 2 can be eliminated by using $0.5\lambda(\text{sum}(W_1 \odot W_1) + \text{sum}(W_2 \odot W_2))$ as regularization strength instead – without a loss of generality – to get $\nabla_{W_1}\mathcal{L} = \nabla_{W_1} MSE + \lambda W_1$ and $\nabla_{W_2}\mathcal{L} = \nabla_{W_2} MSE + \lambda W_2$. In practice, such modification of derivatives to account for the regularization is often handled by the optimization algorithm module in deep learning libraries – so we can avoid explicitly specifying it as part of the loss function. (We will see this later in code.)

- **Dropout:** During training, a fraction of neuron outputs are randomly set to zero, forcing the network to not rely too heavily on any individual neuron. This makes the network more robust. Dropout mode is active during training but should be disabled for evaluation. Dropout was introduced by [Sri+14].

- **Mini-batch gradient descent:** Computing the gradient using small random subsamples of data rows (called *mini-batches*) instead of using the whole data set injects noise into the learning process, helping the training avoid overfitting and getting stuck on suboptimal parameter combinations – and also speeds up the computation. *Stochastic gradient descent* is a variant where a batch consists of just a single observation ($n = 1$) – although occasionally this term is used to describe mini-batch gradient descent as well.

- **Early stopping:** Monitor validation loss during training, and stop training once performance starts degrading (a sign of overfitting).

- **Smaller model / fewer parameters.** A model of lower capacity is less likely to overfit, but at the risk of underfitting if too small.

- **More data.** Ultimately, overfitting is a result of a model learning too much from too little. Larger, more diverse datasets make it harder for the model to overfit and easier to generalize.

Each of these regularization techniques is the subject of substantial research. We will see some of them implemented in code in this and next chapters. For more in-depth coverage, see the encyclopedic reference [GBC16].

5.5 Optimization theory and algorithms for neural nets

Let us now revisit how a neural net actually learns – i.e., how we adjust parameters to minimize the prediction error.

5.5.1 Gradient descent in neural nets

Let w represent the vector of all neural net parameters (flattened), and let $\mathcal{L}(w)$ be our loss function (e.g., MSE + regularization) that we want to minimize.

Regular *gradient descent* iteratively updates weights in the direction that reduces the loss:

$$w_{t+1} = w_t - \gamma \nabla_w \mathcal{L}(w_t).$$

Here:

- w_t is current estimate of parameters, and w_{t+1} is the new estimate.

    - We set initial parameters w_0 randomly.

- γ (pronounced as *gamma*) is a learning rate (a small positive number).

- $\nabla_w \mathcal{L}(w_t)$ is the gradient of the loss with respect to the parameters – that is, an array of partial derivatives of $\mathcal{L}(w)$ with respect to w.

    - In regular gradient descent, $\nabla_w \mathcal{L}(w_t)$ is computed by averaging gradients across all observations in the data (recall that each observation yields a separate gradient estimate). In mini-batch gradient descent, the average is instead taken over a small random sample of data rows – a mini-batch (e.g., 50 observations, although the batch size considered appropriate depends on the data), resulting in a noisier but faster gradient estimate. By the Central Limit Theorem (CLT), gradient estimated on the mini-batch, as a sample mean, will be approximately normally distributed around the true population gradient value.

When the loss function is convex (e.g., quadratic), the gradient descent algorithm comes with theoretical guarantees of convergence to a global minimum.

But for neural networks, the loss landscape is non-convex – potentially filled with saddle points and local minima, yet an empirical fact is that gradient-descent-based algorithms work well even for highly non-linear loss functions. How so?

Though we currently lack a complete definitive theoretical explanation, several insights help:

- In high-dimensional spaces (that is, where loss functions have many parameters) local minima tend to be rare; instead, most critical / stationary points – parameter value combinations that set gradient to zero – seem to be saddle points, with at least some direction in which we can continue to shift parameters values to keep reducing the loss function.

- Well-initialized models combined with stochasticity from mini-batches, in practice, help the optimizer achieve faster convergence; gradient update stochasticity helps optimizer escape shallow local minima or saddle points in the parameter space.

- Modern optimization algorithms, building on the vanilla gradient descent rule above, can adaptively augment their parameter updates based on historical gradient values, which significantly improves training stability.

5.5.2 Adaptive optimizers

Modern deep learning typically uses advanced variants of gradient descent, as the vanilla gradient descent update rule that we have seen so far can lead to unstable training for deep neural networks. By instability, I mean that the loss does not go down consistently and reliably. The core idea behind modern optimizers is to augment the updates for each parameter based on the gradient history throughout training, which tends to lead to smoother training.

Some widely used optimizers include:

- **Adam** and its updated version **AdamW**;
- **RMSprop**;
- **SGD with momentum**;
- **Adagrad**;
- and so on.

All these are available in modern deep learning libraries like PyTorch.

Implementation specifics between these algorithms differ – and, really, do not matter very much, as we can rely on them as black boxes provided by the Python library we use. That said, many of these algorithms are simple in their implementation, so I will demonstrate two of them.

(a) RMSprop

RMSprop algorithm, first release via lecture slides [Hin12], is simple in that it makes a minimal modification to the vanilla gradient descent update, yet it shows great performance in practice. In brief, RMSprop maintains a running average of the squared gradients for each parameter and divides the gradient of each parameter by the square root of its corresponding running average.

The RMSprop algorithm [PyTd]:

- Initialize:

    - Array of neural net parameters w_0.
    - The squared-gradient accumulator $s_0 = 0$ (same shape as w_0).
    - Scalar hyperparameters: learning rate $\gamma = 0.001$, decay rate $\rho = 0.9$ (pronounced *rho*), small constant $\epsilon = 1e - 8$.

- For each step t:

    1. Compute gradient: $g_t \leftarrow \nabla_w \mathcal{L}(w_{t-1})$.
    2. Update gradient magnitude estimate: $s_t \leftarrow \rho\, s_{t-1} + (1 - \rho)\, g_t^2$.
    3. Update parameters using scaled gradient: $w_t \leftarrow w_{t-1} - \gamma \cdot g_t/(\sqrt{s_t} + \epsilon)$.

How it works:

- *Dampens* updates for parameters with large, volatile gradients.
- *Amplifies* updates for parameters with consistently small gradients.

(b) AdamW

AdamW [LH17] is one of the most popular optimizers in deep learning nowadays. It combines two key ideas:

- *Momentum:* Like a ball rolling downhill, we accumulate a moving average of past gradients to smooth out the updates.

- *Adaptive learning rate:* Like RMSprop, we scale updates based on the recent squared gradients to account for gradient magnitude.

In contrast to earlier methods like Adam, which incorporate l_2 regularization (weight decay) directly into the loss function $\mathcal{L}(w)$, AdamW decouples the regularization term from the loss. AdamW applies the regularization-related parameter update separately from the primary objective gradient update – and does not count the regularization gradients in the gradient history.

The AdamW algorithm [PyTa]:

- Initialize:
    - Parameters w_0 and similarly shaped gradient's first and second moment estimates $m_0 = 0$, $v_0 = 0$.
    - Scalar hyperparameters: learning rate γ, decay rates $\beta_1 = 0.9$ and $\beta_2 = 0.999$, small constant $\epsilon = 10^{-8}$, regularization / weight decay strength λ.

- For each step t:
    1. Compute gradient: $g_t \leftarrow \nabla_w \mathcal{L}(w_{t-1})$.
    2. Update first moment: $m_t = \beta_1 m_{t-1} + (1 - \beta_1)g_t$.
    3. Update second moment: $v_t = \beta_2 v_{t-1} + (1 - \beta_2)g_t^2$.
    4. Bias correction: $\hat{m}_t = m_t/(1 - \beta_1^t)$ and $\hat{v}_t = v_t/(1 - \beta_2^t)$.
    5. Update parameters: $w_t \leftarrow w_{t-1} - \gamma \cdot \left(\hat{m}_t/(\sqrt{\hat{v}_t} + \epsilon) + \lambda w_{t-1} \right)$.

The term λw_{t-1} represents the weight decay (l_2 penalty) gradient and is applied outside the main gradient history. This means regularization gradient does not get recorded in the moment estimates m and v, which remain purely focused on the gradients of the primary loss. This decoupling of primary objective vs. regularization (weight decay) gradients distinguishes AdamW from its predecessor Adam.

> **i** AdamW in PyTorch
>
> In PyTorch, you can invoke AdamW like this:
> ```
> optimizer = torch.optim.AdamW(model.parameters(),
> lr=0.001, weight_decay=0.01)
> ```

5.6 Python code implementation of neural nets

Here we will implement neural net training on a toy data set in Python – first, from scratch in NumPy and, second, using the dedicated PyTorch library.

5.6.1 Loading data

We will use here life expectancy data from the CIA Fact Book website as an example: https://www.cia.gov/the-world-factbook/references/guide-to-country-comparisons/.

This snippet of code scrapes the data from the web and saves it in a .csv format – the data table is identified in HTML using a CSS selector obtained by examining the site's raw HTML (therefore the code depends on the website structure, so could be affected if the site changes):

```python
import pandas as pd
import requests # url access
from bs4 import BeautifulSoup # html parsing
from io import StringIO # for pandas to accept html

var_names = ["life-expectancy-at-birth",
"real-gdp-per-capita",
"obesity-adult-prevalence-rate",
"alcohol-consumption-per-capita"]

url_template = "https://www.cia.gov/the-world-factbook/field/
 ↪ {}/country-comparison/"
```

```python
# web scraping
data_list = []
for var in var_names:
    url = url_template.format(var)
    response = requests.get(url)
    soup = BeautifulSoup(response.text, "html.parser")
    table = soup.select_one(".content-table") # css selector
    df = pd.read_html(StringIO(str(table)))[0]
    df = df.rename(columns={df.columns[2]: var})
    data_list.append(df.iloc[:,[1,2]]) # storing country and
 ↪   metric cols

# merging
data = pd.concat([df.set_index('Country') for df in
 ↪   data_list],
                 axis=1, join='inner').dropna().reset_index()

# converting GDP string to float
data.loc[:,'real-gdp-per-capita'] =
 ↪   data['real-gdp-per-capita'
     ].str.replace("$","").str.replace(",","").astype(float)

data.to_csv('life_expectancy_data.csv', index=False)
```

Load the data from the file:

```python
import pandas as pd
import numpy as np
import matplotlib.pyplot as plt

data = pd.read_csv('./life_expectancy_data.csv')

data.columns
```

```
Index(['Country', 'life-expectancy-at-birth', 'real-gdp-per-
capita',
```

```
      'obesity-adult-prevalence-rate', 'alcohol-consumption-
per-capita'],
      dtype='object')
```

Select desired columns:

```
data['real-gdp-per-capita'] =
↪  np.log(data['real-gdp-per-capita'])  # taking log GDP

print(data.shape)
print(data.head(3).T)
```

```
(188, 5)
                                         0          1          2
Country                          Singapore      Japan     Canada
life-expectancy-at-birth              86.7       85.2       84.2
real-gdp-per-capita              11.795092  10.738568  10.945529
obesity-adult-prevalence-rate          6.1        4.3       29.4
alcohol-consumption-per-capita        1.81       8.36        8.0
```

Notice that we log-transformed the GDP variable. This reduces the skewness in the distribution of GDP values and brings their magnitudes to a more manageable scale, helping avoid possible numerical instability caused by large gradient values during optimization.

Prepare NumPy arrays:

```
X = data.iloc[:, 2:].values
y = data.iloc[:, 1].values

print("Input data X shape:", X.shape)
print("Prediction target y shape:",y.shape)
```

```
Input data X shape: (188, 3)
Prediction target y shape: (188,)
```

Reshape predicted life expectancy vector into explicit 2D matrix:

```
y = y[:,np.newaxis] # reshaping
print("Prediction target y shape:",y.shape)
```

```
Prediction target y shape: (188, 1)
```

5.6.2 Neural net training from scratch with NumPy

Some code organization ideas in the following NumPy neural net implementation, like storing parameter and gradient arrays as dictionaries and the arrangement of the forward and backward pass into a single function, are inspired by the classic Stanford University's CS231n course assignments (https://cs231n.stanford.edu/).

We will first define the neural net structure that will hold all the weights and the partial derivatives of MSE with respect to those weights.

We can initialize weights of desired shape like this:

```
np.random.randn(4,3)
```

```
array([[-0.2257763 ,  0.0675282 , -1.42474819],
       [-0.54438272,  0.11092259, -1.15099358],
       [ 0.37569802, -0.60063869, -0.29169375],
       [-0.60170661,  1.85227818, -0.01349722]])
```

The following function will initialize neural net as a dictionary that stores the weights across two layers of neurons (hidden layer and output layer). Weights that multiply inputs are initialized as random normal. Bias parameters are initialized to zero.

```
def initialize_net(input_size, hidden_size, output_size):
    np.random.seed(999)
    net = {}
    net['W1'] = np.random.randn(input_size, hidden_size)
```

```
net['b1'] = np.zeros((1, hidden_size))
net['W2'] = np.random.randn(hidden_size, output_size)
net['b2'] = np.zeros((1, output_size))
return net
```

We initialize the net with 3 inputs, 4 hidden layer neurons, and 1 output neuron - this is the net we have used as a running example throughout this chapter:

```
net = initialize_net(3, 4, 1)
net
```

```
{'W1': array([[ 0.12715784,  1.40189088,  0.31481499, -
0.85844916],
        [-0.26613444, -0.64890071,  1.56626757, -2.09137019],
        [ 1.45632806,  0.94529342, -0.40020119,  0.3152273 ]]),
 'b1': array([[0., 0., 0., 0.]]),
 'W2': array([[-1.11006083],
        [-0.58482153],
        [-0.18840956],
        [ 0.81302365]]),
 'b2': array([[0.]])}
```

Forward pass

We unpack parameters from the neural net dictionary into variables for simplicity and compute forward pass.

```
# unpacking
W1, b1, W2, b2 = net['W1'], net['b1'], net['W2'], net['b2']
N, D = X.shape

# non-linearity function - elementwise maximum
relu = lambda x: np.maximum(x, 0)

# forward pass - for input data X
```

```
Z = relu(X.dot(W1) + b1)
y_hat = Z.dot(W2) + b2
print("Prediction shape:", y_hat.shape)

# MSE between prediction and data y
pred_error = y_hat - y
print("Prediction error (residual) shape:", pred_error.shape)

print("MSE:", np.mean(pred_error**2))
print("MSE (using dot product):", np.dot(pred_error.T,
 ↪  pred_error)/N )
```

```
Prediction shape: (188, 1)
Prediction error (residual) shape: (188, 1)
MSE: 7931.881911375143
MSE (using dot product): [[7931.88191138]]
```

```
# full loss accounting for regularization

# mse
loss = np.dot(pred_error.T, pred_error) / N

# regularization (optimizing to shrink parameter magnitude
 ↪  too)
reg = 0.01
loss += 0.5 * reg * (np.sum(W1*W1) + np.sum(W2*W2))
```

Backward pass

We now implement the computation of partial derivatives of our loss function with respect to each parameter in the net – that is, gradients. We set up a dictionary to hold the gradients for the parameters. Note that parameter gradient arrays should be the same shape as the parameter arrays themselves. Function `np.zeros_like(x)` gives an array of zeros shaped like target array x. Here is the computation for our net:

```python
grads = {}

grads['W2'] = np.zeros_like(W2)
grads['b2'] = np.zeros_like(b2)
grads['W1'] = np.zeros_like(W1)
grads['b1'] = np.zeros_like(b1)

dMSE_dy_hat = 2*(y_hat - y)

grads['W2'] += Z.T.dot(dMSE_dy_hat) / N
grads['W2'] += reg * W2  # regularization for non-intercept
↪ parameters

grads['b2'] += np.ones((1,N)).dot(dMSE_dy_hat) / N

dMSE_dZ = dMSE_dy_hat.dot(W2.T)

grads['W1'] += X.T.dot((Z>0) * dMSE_dZ) / N
grads['W1'] += reg * W1

grads['b1'] += np.ones((1,N)).dot((Z>0) * dMSE_dZ) / N

for k,v in grads.items():
    print(k+":\n", np.round(v,3))
```

```
W2:
 [[-7.922770e+02]
 [-1.186301e+03]
 [-5.682298e+03]
 [ 8.000000e-03]]
b2:
 [[-176.732]]
W1:
 [[ 1.412741e+03  8.333940e+02  3.215070e+02 -9.000000e-03]
 [ 2.569732e+03  1.463900e+03  6.626120e+02 -2.100000e-02]
 [ 9.501560e+02  5.029160e+02  1.709570e+02  3.000000e-03]]
b1:
```

```
[[143.929  85.768  33.298   0.   ]]
```

As you can see, it does not look too terrible in code, but it is error-prone to write down and we really should let the software do these computations automatically.

Also notice that we are adding the computed derivatives to zero value arrays instead of overwriting them. This ensures we correctly accumulate derivatives from diffident portions of the loss function – MSE and regularization – as well as serves as a check to ensure gradients are of the right shape (the shape of their corresponding parameter arrays).

We now put these together into a single wrapper function that return prediction if no target is provided and returns total loss, MSE, and gradients if prediction target is given.

```python
def run_net(X, net, y=None, reg=0.01):

    W1, b1, W2, b2 = net['W1'], net['b1'], net['W2'],
↪   net['b2']
    N, D = X.shape

    relu = lambda x: np.maximum(x, 0)

    # forward pass - computing prediction
    Z = relu(X.dot(W1) + b1)
    y_hat = Z.dot(W2) + b2

    # if we do not provide prediction target y
    # simply do a forward pass and return prediction y_hat
    if y is None:
        return y_hat

    # loss function
    pred_error = y_hat - y
    MSE = np.dot(pred_error.T, pred_error) / N

    # L2 regularization
    loss = MSE + 0.5 * reg * (np.sum(W1*W1) + np.sum(W2*W2))
```

```python
# backward pass - computing gradients
grads = {}

grads['W2'] = np.zeros_like(W2)
grads['b2'] = np.zeros_like(b2)
grads['W1'] = np.zeros_like(W1)
grads['b1'] = np.zeros_like(b1)

dMSE_dy_hat = 2*(y_hat - y)

grads['W2'] += Z.T.dot(dMSE_dy_hat) / N
grads['W2'] += reg * W2
grads['b2'] += np.ones((1,N)).dot(dMSE_dy_hat) / N

dMSE_dZ = dMSE_dy_hat.dot(W2.T)

grads['W1'] += X.T.dot((Z>0) * dMSE_dZ) / N
grads['W1'] += reg * W1
grads['b1'] += np.ones((1,N)).dot((Z>0) * dMSE_dZ) / N


return loss.item(), MSE.item(), grads
```

Notice that we can get parameter names like this.

```python
loss, MSE, grads = run_net(X, net, y=y)
print("Parameter arrays:", grads.keys())
print("MSE:", MSE)
```

```
Parameter arrays: dict_keys(['W2', 'b2', 'W1', 'b1'])
MSE: 7931.881911375143
```

We now implement a training loop using gradient descent, keeping track of losses for visualization.

```
losses = []
MSEs = []
learning_rate = 0.0001
for i in range(100000):
    loss, MSE, grads = run_net(X, net, y=y)
    losses.append(loss)
    MSEs.append(MSE)
    # parameter updating - gradient descent
    names = grads.keys()
    for j in names:
        net[j] -= grads[j] * learning_rate


print("Initial loss", losses[0])
print("Final loss", losses[-1])
```

```
Initial loss 7931.960317329429
Final loss 15.667574456849723
```

```
print("Initial MSE", MSEs[0])
print("Final MSE", MSEs[-1])
```

```
Initial MSE 7931.881911375143
Final MSE 15.527824145612641
```

Plotting the history of losses (MSE + regularization) through training iterations.

```
plt.plot(losses[200:])
plt.ylabel("Total loss (MSE + reg.)")
plt.xlabel("Iteration")
plt.xticks(rotation=45)
plt.show()
```

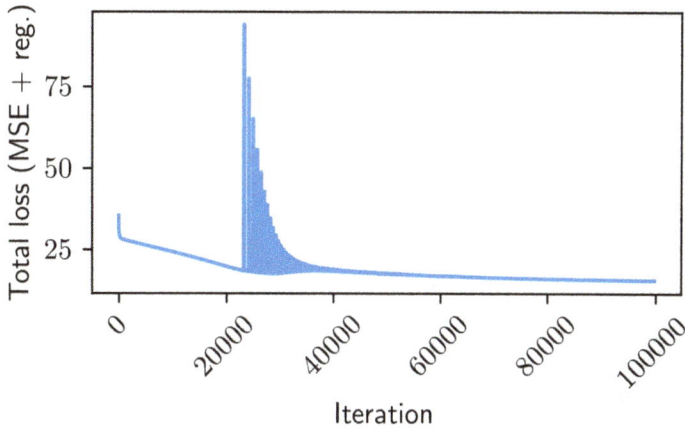

Figure 5.5: History of loss (MSE + regularization) through iterations

The ragged appearance of the loss history plot indicates some updates led to (sharp!) loss increases rather than decreases – which could be ameliorated by taking smaller gradient steps (that is, using smaller learning rate). However, we will see soon how adaptive optimizers enable more stable training, compared to vanilla gradient descent, at the same learning rate.

5.6.3 Linear regression benchmark

Notice that simple linear regression achieves higher (worse) MSE at this prediction task, compared to the neural net – so a non-linear prediction function seems to help here.

```
Xn = np.append( np.ones(X.shape[0])[:,np.newaxis],  X,  1)
beta = np.linalg.inv(Xn.T.dot(Xn)).dot(Xn.T.dot(y))
print('Linear regression MSE:', np.mean((y-Xn.dot(beta))**2))
```

```
Linear regression MSE: 15.888872357934279
```

```
print("Neural net MSE", MSEs[-1])
```

```
Neural net MSE 15.527824145612641
```

5.6.4 RMSprop updates

Let us try the same neural net training with RMSprop:

```
# reset net
net = initialize_net(3, 4, 1)
s = {} # squared gradient accumulator
for j in net.keys():
    s[j] = np.zeros_like(net[j])
losses = []
MSEs = []
learning_rate = 0.0001
rho = 0.9
eps = 1e-8
for i in range(100000):
    loss, MSE, grads = run_net(X, net, y=y)
    losses.append(loss)
    MSEs.append(MSE)
    # update gradient magnitude estimate
    for j in net.keys():
        s[j] = rho*s[j] + (1-rho)*grads[j]**2
    # parameter updating with RMSprop
    for j in net.keys():
        net[j] -= grads[j]*learning_rate/(s[j]**0.5+eps)

print("Initial MSE", MSEs[0])
print("Final MSE", MSEs[-1])


Initial MSE 7931.881911375143
Final MSE 13.852495772282504
```

Plotting the history of loss through training iterations:

```
plt.plot(losses[200:])
plt.ylabel("Total loss (MSE + reg.)")
plt.xlabel("Iteration")
plt.xticks(rotation=45)
plt.show()
```

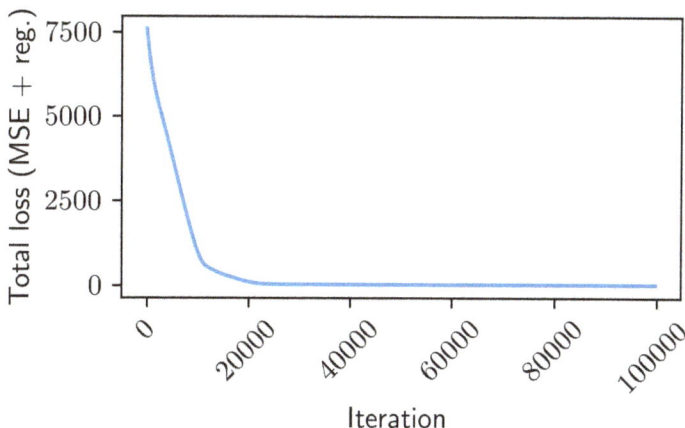

Figure 5.6: History of loss (MSE + regularization) through iterations (RM-Sprop)

Note the absence of instability in loss evolution with RMSprop. Also, RMSprop MSE value is lower than what we got with vanilla gradient descent!

5.6.5 Neural net MSE loss surface

Let us see how the neural net loss (MSE + regularization) – that we aim to minimize – looks when we tweak a single parameter in each W1 and W2 arrays of weights, for a randomly initialized net, keeping other parameters fixed:

```python
from mpl_toolkits.mplot3d import Axes3D
from matplotlib.colors import LogNorm

# duplicate net learned via RMSprop
net = initialize_net(3, 4, 1)

# neural net loss function
def mse_mod(wa, wb):
    net['W1'][0,0] = wa
```

```
    net['W2'][0,0] = wb
    loss, MSE, _ = run_net(X, net, y=y)
    return loss

# create a meshgrid around learned weights
a = np.linspace(-6, 6, 200)
b = np.linspace(-6, 6, 200)
A, B = np.meshgrid(a, b)

# compute function values
Z = np.zeros_like(A)
for i in range(A.shape[0]):
    for j in range(A.shape[1]):
        Z[i, j] = mse_mod(A[i, j], B[i, j])

# neural net loss plot
fig, ax = plt.subplots(1, 1, figsize=(5, 5),
 ↪  subplot_kw=dict(projection='3d'))
surf = ax.plot_surface(A, B, Z, alpha=0.8, cmap='jet',
 ↪  norm=LogNorm())
ax.set_xlabel("W1[0,0]")
ax.set_ylabel("W2[0,0]")
ax.set_zlabel("Loss", labelpad=8)
fig.colorbar(surf, ax=ax, shrink=0.5, aspect=10, pad=0.2)
ax.view_init(elev=50, azim=-30)
plt.show()
```

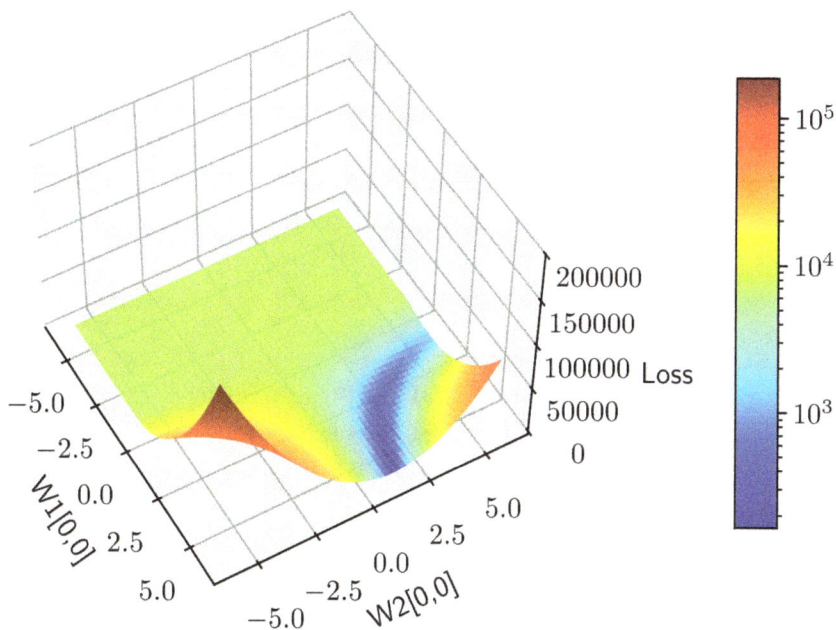

Figure 5.7: Neural net loss surface (MSE + regularization) in 3D

In Figure 5.7, you can see the bending ravine in the loss landscape indicating the low-loss area of the parameter space, where the gradient descent might take us. Of course, the loss varies with respect to all the involved parameters, but visualizing the loss landscape in more than 3D is not something we can handle very well.

5.6.6 PyTorch implementation and train / test split

PyTorch is currently the leading deep learning library in practice. It was originally developed by Meta (formerly Facebook) as a Python-based successor to the Torch library, which was built in the Lua programming language. Today, PyTorch has become the de facto standard for research and production in deep learning.

There are alternative libraries, for example:

- TensorFlow (developed by Google) was once dominant but has lost traction in recent years [Red], in part due to early architectural decisions that made it less intuitive and harder to debug.

- JAX (also by Google) is a newer library that emphasizes composability and performance, and is popular in academic settings. It offers a clean design and excellent automatic differentiation – but still lacks the maturity and ecosystem breadth of PyTorch.

> 💡 Tip
>
> Unless you have a specific technical need that another library satisfies better, stick with PyTorch. It strikes the best balance between usability, flexibility, performance, and community support.

Load PyTorch (install it first if necessary):

```
# to install pytorch, uncomment and run:
# !conda install pytorch torchvision -c pytorch

import torch
import torch.nn.functional as F
```

> ℹ PyTorch installation
>
> The basic installation ensures PyTorch runs on your CPU – and potentially on Apple's MPS back-end if you're using a Mac (learn more about MPS [App]). To use PyTorch with a GPU, you must first install the appropriate GPU drivers (e.g., CUDA for NVIDIA). Once a compatible driver is installed, follow one of the setup commands provided on the official PyTorch site [PyTb] to install a GPU-enabled version of PyTorch. Keep in mind – GPU setup can be complex and may require careful alignment between the driver and PyTorch versions.

Setting random seeds so everything is reproducible. PyTorch library requires its own random seed being set.

```
torch.manual_seed(1234)
np.random.seed(1234)
```

> **i** Reproducibility
>
> Some operations, when it comes to deep learning, especially if one
> runs code on GPUs, may be inherently non-deterministic – for exam-
> ple, different ordering of limited-precision addition leads to different
> rounding and thus to different numerical results; parallel computation
> of gradients on different data samples and updates using those may
> lead to varied order of updates, etc. – so full reproducibility may not
> be achievable on industrial-scale training. However, if it is strictly
> required, PyTorch offers some facilities to try to support full repro-
> ducibility, such as `torch.use_deterministic_algorithms(True)`
> flag, usually, at the cost of time – because deterministic operations
> tend to be slower [PyTc].

We will now randomly split the data into the train and test sets. We will use
the train set to estimate gradients and update the model parameters, and
we will use the test set to monitor the model's hold-out performance. The
hold-out evaluation gives us an idea of how the model would work on new
incoming data – as opposed to the data that the model has had a chance to
memorize / overfit because it was trained on it.

```
ind = list(range(X.shape[0]))
np.random.shuffle(ind)
n = int(X.shape[0]*0.6)
idx_train, idx_test = ind[:n], ind[n:]

X_train = X[idx_train]
y_train = y[idx_train]

X_test = X[idx_test]
y_test = y[idx_test]
```

> ⚠ Information leakage
>
> A word of caution regarding the train-test split – when transforming the data in addition to training a model, it is essential to avoid *data leakage*. For example, it is common to perform *normalization* – standardizing a variable by subtracting its mean and dividing by its standard deviation – so all variables are on the same scale, which can help with training stability. A common mistake is to normalize the entire data set *before* splitting it into training and test sets. This allows information from the test set to influence the transformation applied to the training data, which can artificially inflate the model's test-set performance, as the test set is no longer truly "hold-out." Instead, a conceptually clean approach is to compute the mean and standard deviation on the train set only and then use those to normalize both train and test sets. This ensures the test data remains truly unseen during training. That said, transformations like log-scaling do not involve parameters learned from the data and can be safely applied before data splitting without causing information leakage.

Transforming NumPy arrays to PyTorch tensors:

```
X_train = torch.from_numpy(X_train).float()
y_train = torch.from_numpy(y_train).float()
X_test = torch.from_numpy(X_test).float()
y_test = torch.from_numpy(y_test).float()
```

Defining our neural net structure:

```
net = torch.nn.Sequential(
    torch.nn.Linear(3,4),
    torch.nn.ReLU(),
    torch.nn.Linear(4,1)
)
```

We will be using the RMSprop gradient descent algorithm:

```
optimizer = torch.optim.RMSprop(
    net.parameters(), lr=0.0001, weight_decay=0.01)
```

Loss function:

```
mse = torch.nn.MSELoss()
```

> **i** Weight decay means l_2 regularization
>
> Ridge regularization is applied implicitly via the weight decay parameter in the optimizer – so loss is just the MSE.

Training loop:

```
losses = []
for i in range(100000):
    y_train_pred = net(X_train)
    train_loss = mse(y_train_pred, y_train)
    optimizer.zero_grad() # zero out gradient storage
    train_loss.backward() # compute gradients
    optimizer.step()      # update parameters using gradients

    # loss tracking code - evaluate every 10000 steps
    if ((i+1) % 10000 == 0):
        # disable gradient tracking inside the scope
        with torch.no_grad():
            y_test_pred = net(X_test)
            test_loss = mse(y_test_pred, y_test)
            losses.append({
                "Train": train_loss.item(),
                "Test": test_loss.item(),
                "Iter": i,
                })
        print(f"Step {i}: "
            f"Train loss = {train_loss.item():.4f}, "
            f"Test loss = {test_loss.item():.4f}")
```

```
Step 9999: Train loss = 814.5470, Test loss = 828.2895
Step 19999: Train loss = 209.6220, Test loss = 219.0393
Step 29999: Train loss = 17.0300, Test loss = 36.7629
Step 39999: Train loss = 12.8890, Test loss = 29.1695
Step 49999: Train loss = 11.4556, Test loss = 24.7408
Step 59999: Train loss = 11.3552, Test loss = 23.9127
Step 69999: Train loss = 11.3547, Test loss = 23.9036
Step 79999: Train loss = 11.3545, Test loss = 23.8974
Step 89999: Train loss = 11.3541, Test loss = 23.8902
Step 99999: Train loss = 11.3539, Test loss = 23.8833
```

```
losses = pd.DataFrame(losses)
plt.plot(losses['Iter'],losses['Train'],label='Train')
plt.plot(losses['Iter'],losses['Test'],label='Test',ls=":")
plt.ylabel('MSE')
plt.xlabel('Iteration')
plt.xticks(rotation=45)
plt.legend()
plt.show()
```

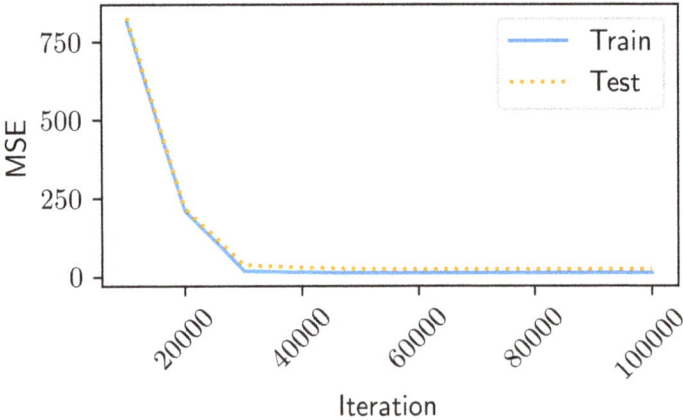

Figure 5.8: History of MSE through iterations

> ⚠ Train and evaluation mode
>
> For more complex neural nets that include features like dropout, these mechanisms must be enabled during training and disabled during evaluation. In PyTorch, e.g., for our model `net`, this would be done by calling `net.train()` before training and `net.eval()` before computing the loss. For our simple model this would make no difference, as it contains no advanced features to turn on and off, but we will rely on these functions when working with language models in the next chapter.

```
# final evaluation (we could get it from losses dataframe
  ↳ too)
with torch.no_grad():
    print('Train set MSE:',mse(net(X_train), y_train).item())
    print('Test set MSE:',mse(net(X_test), y_test).item())
```

```
Train set MSE: 11.35384464263916
Test set MSE: 23.883264541625977
```

Final train set error is lower than the test set error – so, perhaps, we overfit the train set and need to increase the regularization strength. Although this could also be just a statistical aberration because the data set we use here as an example is relatively small and we could just be unlucky with some large outlier value in the test set.

Overall, we can see how much simpler everything is when using PyTorch – we only define forward computation and the library does the rest. No need to manually compute gradients.

It may be useful to extract NumPy arrays from PyTorch tensors. In recent PyTorch versions, it can be achieved via `.numpy(force=True)` command like this:

```
y_test_pred = net(X_test).numpy(force=True)
```

It handles all kinds of scenarios, including tensor data being located on the GPU instead of the CPU. See https://docs.pytorch.org/docs/stable/gener ated/torch.Tensor.numpy.html for more details about what happens under the hood of this command.

5.7 Bootstrap for model evaluation

Noting that test set MSE for the trained neural net is itself a random variable, we can get bootstrap confidence intervals (CIs) on the test MSE metric:

```python
torch.manual_seed(1234)

def bootstrap_mse(X, y):
    n = X.shape[0]
    ind = torch.randint(0, n, (n,))
    X_b, y_b = X[ind], y[ind] # bootstrap sample
    with torch.no_grad():
        y_b_pred = net(X_b)
        out = mse(y_b_pred, y_b).item()
    return out

mse_b_estimates = []
for i in range(1000):
    mse_b_estimates.append(
        bootstrap_mse(X_test, y_test))

print(f"Median test MSE: {np.median(mse_b_estimates):.2f}")
print("CI 95% (2.5th, 97.5th percentile): "
      f"{np.percentile(mse_b_estimates, 2.5):.2f}, "
      f"{np.percentile(mse_b_estimates,97.5):.2f}")
```

```
Median test MSE: 23.83
CI 95% (2.5th, 97.5th percentile): 15.88, 33.34
```

There is quite a bit of variance in the test MSE!

The fact that the train set MSE point estimate we obtained earlier lies below the lower edge of this 95% test MSE confidence interval (CI) suggests we might be observing real overfitting rather than just a statistical fluke.

We can also use such bootstrap 95% test MSE confidence intervals to compare models. However, note that these bootstrap intervals describe the performance of a particular trained model – they reflect the variability from test set sampling, but not the additional variability we would see if we retrained the model many times (e.g., from different initializations or data shuffles). Such training-related variability may or may not matter – it is crucial if we are comparing architectures rather than specific model instances, especially when training is unstable. Methods like cross-validation, where a model is repeatedly trained and evaluated across multiple data splits, can capture both these sources of variability, but such methods are also more computationally intensive because they require retraining the model from scratch.

5.8 Figuring out the right neural net architecture

Of course, we are not limited to the simple, single hidden-layer neural net used as an example in this chapter. In practice, neural networks can have multiple hidden layers, each adding more representational power to the model, as depicted in Figure 5.9.

Neurons also do not need to connect to every output of the preceding layer. We can define almost arbitrarily complex (and sometimes quite unusual) connection patterns. For example, in *convolutional neural networks (CNNs)*, often used in image processing, a neuron may only connect to a small patch of the input image – focusing on local features rather than the whole.

So, what is the right neural network architecture?

There is no universal answer – and designing effective architectures is often as much art as science. It typically involves a great deal of empirical testing and experimentation. A substantial body of research (and many PhD hours) has been devoted to discovering architectures that work well in different domains.

In the next chapter, we will examine one particularly influential architecture – the **transformer** – which underlies modern large language models and has played a major role in their success.

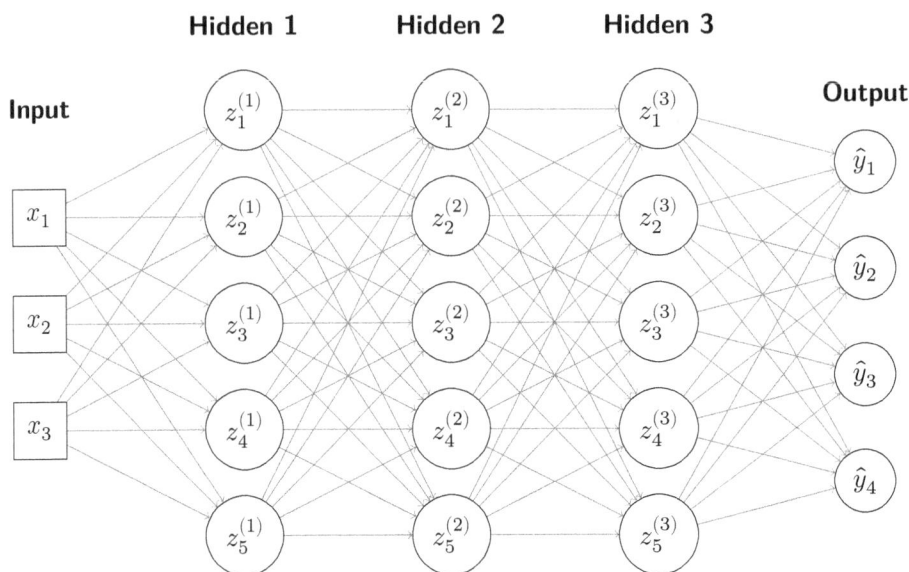

Figure 5.9: A more complex neural net illustration

5.9 Role of GPUs

Neural network training and inference (a fancy name for using the neural net to make predictions) are essentially sequences of matrix operations – lots of them. It turns out that graphics processing units (GPUs), originally developed to accelerate video game graphics (which also rely heavily on matrix math), are extremely well-suited for this kind of computation.

Compared to standard CPUs, GPUs can offer orders-of-magnitude speedups, especially for large models and datasets. Using a single GPU in place of a CPU can be the difference between training a model in *hours vs. months*.

This is a major reason why *NVIDIA* – a company specializing in GPUs – has become central to the current AI boom – and why its stock price has been in the news. Large language models are essentially matrix factories.

5.10 Deep learning history

Neural networks have been around since the 1950s, originally under the name *perceptron* due to Cornell University psychologist Frank Rosenblatt. But they remained on the fringes of mainstream machine learning for decades. What triggered their dramatic resurgence in the 2000s?

Three key developments:

- Large-scale datasets became more available (e.g., ImageNet [Den+09]).

- Improved optimization algorithms (like RMSprop and Adam) enabled stable training of deep models.

- More powerful hardware (especially GPUs) made large-scale training feasible.

Another factor was branding. Rebranding the field of neural networks as *deep learning* starting in 2000s helped reframe it as something new and exciting – and gave it the visibility and momentum it needed.

> **i** Deep learning brand
>
> As a humorous, historical aside, consider Richard Bellman's comment in his autobiography *The Eye of the Hurricane* (1984) about coining the term *dynamic programming*: "Where did the name, dynamic programming, come from? ... It also has a very interesting property as an adjective, and that is its impossible to use the word, dynamic, in a pejorative sense. ... Thus, I thought dynamic programming was a good name. It was something not even a Congressman could object to. So I used it as an umbrella for my activities."

Since then, deep learning has delivered ground-breaking models across such diverse areas as image recognition, image and text generation, autonomous driving, protein folding, weather forecasting, etc. – all based on the simple foundations laid out in this chapter.

See [Fra20; Sch15] for more deep learning history.

5.11 Advanced: Automatic differentiation

For the curious, here is a very basic example of how a *reverse-mode* (i.e.,
backpropagation) automatic differentiation engine can be written in Python.
It performs a forward pass, constructing a computational graph, and then
performs backpropagation using the chain rule, ensuring it is done in the
correct order based on the computational graph. This is minimal example,
but could be extended, with enough work, to handle the full neural net
computation.

```python
# A variable is a dict that stores:
# - value: the current numeric value
# - grad: its gradient (starts at 0, filled in later)
# - prev: inputs used to compute the variable
# - backward: function to compute its local gradients
def create_var(value):
    return {'value': value, 'grad': 0.0, 'prev': [],
     ↪  'backward': lambda: None}

# define addition operation with autodiff tracking
def add(a, b):
    out = create_var(a['value'] + b['value'])
    out['prev'] = [a, b] # added variables

    # how the gradient flows backward through addition
    # the gradient with respect to a and b is just 1 * the
    ↪  output gradient
    # (d(a+b)/da = 1)
    def _backward():
        a['grad'] += 1.0 * out['grad']
        b['grad'] += 1.0 * out['grad']
    out['backward'] = _backward
    return out

# define multiplication operation with autodiff tracking
def mul(a, b):
    out = create_var(a['value'] * b['value'])
    out['prev'] = [a, b]
```

```
        # how the gradient flows backward through multiplication
        ↪  (chain rule)
        # (d(a*b)/da = b)
        def _backward():
            a['grad'] += b['value'] * out['grad']
            b['grad'] += a['value'] * out['grad']
        out['backward'] = _backward
        return out

# ReLU operation
def relu(a):
    out_val = max(0.0, a['value'])
    out = create_var(out_val)
    out['prev'] = [a]
    def _backward():
        a['grad'] += (1.0 if a['value'] > 0 else 0.0) *
↪  out['grad']
    out['backward'] = _backward
    return out

# topological sort: ensures we do the backward pass in the
↪  correct order
def topological_sort(var, visited=None, order=None):
    if visited is None: visited = set()
    if order is None: order = []
    if id(var) not in visited:
        visited.add(id(var))
        for prev in var['prev']:
            topological_sort(prev, visited, order)
        order.append(var)
    return order

# backward pass: initializes gradient of output to 1, and
↪  walks the graph
def backward(var):
    var['grad'] = 1.0  # seed gradient at output
```

```
for node in reversed(topological_sort(var)):
    node['backward']()
```

Topological sort: To correctly apply the chain rule during backpropagation, we must compute gradients in a specific order – from the final output back to the inputs. This is the reverse of the order in which the forward computations were performed. To ensure this, we perform a topological sort of the computation graph. Each intermediate result in the computation (a "node") depends on earlier values – these dependencies are stored in the `prev` field of each node. During the topological sort, we recursively visit each node's inputs before the node itself, ensuring that no node is processed before all of its dependencies have been handled. Once the nodes are sorted in this dependency-respecting order, we iterate over them in reverse. This guarantees that when we call a node's `backward()` function, the gradients of all outputs that depend on it have already been computed, allowing the correct accumulation of partial derivatives backward through the graph.

Example:
$$f(x) = \text{ReLU}(x^2 + 3x)$$

at $x = 2$.

```
x = create_var(2.0)
x2 = mul(x, x)                 # x^2
x3 = mul(x, create_var(3.0))   # 3x
y = add(x2, x3)                # x^2 + 3x
z = relu(y)                    # ReLU(x^2 + 3x)

backward(z)

print("f(x) =", z['value'])      # should be 10.0
print("df/dx(x) =", x['grad'])   # should be 7.0 if x^2 + 3x >
 ↳  0


f(x) = 10.0
df/dx(x) = 7.0
```

Other operations can be implemented in a similar fashion – such as division, power, exp, log, etc., allowing automated gradient computation for extremely large and complex computational graphs.

Alternatives to reverse-mode automatic differentiation also exist, for example, automatic differentiation based on finite differences (numeric approximation of the gradient), an example of which we saw in the previous chapter.

You can read more about automatic differentiation here [Bay+18].

5.12 Discussion

In this chapter, we explored how to build and train simple neural networks. To make a large language model (LLM), we simply apply this technology to the task of predicting text sequences. We will see how to do it next.

5.13 Further learning resources

- Deep learning:
  - Encyclopedic reference: [GBC16].
  - Stanford CS231n course: https://cs231n.stanford.edu/.
- Automatic differentiation: [Bay+18] and (advanced) [GW08].

5.14 References

[App]   Apple. *Accelerated PyTorch training on Mac.* URL: https://de
            veloper.apple.com/metal/pytorch (visited on 05/05/2025).

[Bay+18]   Atilim Gunes Baydin et al. *Automatic differentiation in ma-
            chine learning: A survey.* 2018. arXiv: 1502.05767 [cs.SC] .
            URL: https://arxiv.org/abs/1502.05767.

[Den+09]   Jia Deng et al. "ImageNet: A large-scale hierarchical image
            database". In: *IEEE CVPR.* Ieee. 2009, pp. 248–255.

[Fra20]   Alexander L Fradkov. "Early history of machine learning". In:
            IFAC-PapersOnLine 53.2 (2020), pp. 1385–1390.

[GBC16]   Ian Goodfellow, Yoshua Bengio, and Aaron Courville. *Deep Learning*. http://www.deeplearningbook.org. MIT Press, 2016.

[GW08]   Andreas Griewank and Andrea Walther. *Evaluating Derivatives: Principles and Techniques of Algorithmic Differentiation*. SIAM, 2008.

[Hin12]   Geoffrey Hinton. "Lecture 6, Neural Networks for Machine Learning". In: *University of Toronto* (2012). URL: https://www.cs.toronto.edu/~tijmen/csc321/slides/lecture_slides_lec6.pdf.

[LH17]   Ilya Loshchilov and Frank Hutter. "Decoupled weight decay regularization". In: *arXiv preprint arXiv:1711.05101* (2017).

[Mad+19]   Wesley J Maddox et al. "A simple baseline for Bayesian uncertainty in deep learning". In: *NeurIPS* 32 (2019).

[PyTa]   PyTorch. *AdamW*. URL: https://docs.pytorch.org/docs/stable/generated/torch.optim.AdamW.html (visited on 05/30/2025).

[PyTb]   PyTorch. *Install PyTorch locally*. URL: https://pytorch.org/get-started/locally/ (visited on 05/05/2025).

[PyTc]   PyTorch. *Reproducibility*. URL: https://pytorch.org/docs/stable/notes/randomness.html (visited on 05/05/2025).

[PyTd]   PyTorch. *RMSprop*. URL: https://docs.pytorch.org/docs/stable/generated/torch.optim.RMSprop.html (visited on 05/30/2025).

[Red]   Reddit. *The Downfall of TensorFlow*. URL: https://www.reddit.com/r/MachineLearning/comments/11r363i/d_2022_state_of_competitive_ml_the_downfall_of/ (visited on 05/30/2025).

[Sch15]   Jürgen Schmidhuber. "Deep learning in neural networks: An overview". In: *Neural Networks* 61 (2015), pp. 85–117.

[Sri+14]   Nitish Srivastava et al. "Dropout: A Simple Way to Prevent Neural Networks from Overfitting". In: *Journal of Machine Learning Research* 15.56 (2014), pp. 1929–1958.

[Wika]   Wikipedia. *Automatic differentiation*. URL: https://en.wikipedia.org/wiki/Automatic_differentiation (visited on 05/05/2025).

[Wikb]   Wikipedia. *Rectifier (neural networks)*. URL: https://en.wikipedia.org/wiki/Rectifier_(neural_networks) (visited on 05/30/2025).

[Wikc]   Wikipedia. *Subderivative*. URL: https://en.wikipedia.org/wiki/Subderivative (visited on 05/30/2025).

[Wikd]   Wikipedia. *Universal approximation theorem*. URL: https://en.wikipedia.org/wiki/Universal_approximation_theorem (visited on 05/05/2025).

6 Language models for text prediction

In this chapter, we explore how to represent text numerically and implement the transformer neural network architecture that has achieved great success at generating human-like text by predicting and sampling new text based on past text.

Before we begin, please ensure you have previously interacted – via text – with a commercial-grade large language model (LLM) such as OpenAI's ChatGPT, Anthropic's Claude, Google's Gemini, Meta's Llama, DeepSeek, or a similar system – so you are familiar with what an LLM user experiences (you can find these ready to interact with through online search).

In a typical session, you enter plain text into a chat window, the LLM processes that text, and then returns its own text response. Any formatting you see (bold, italics, lists, etc.) is simply a rendering of raw markup instructions embedded in the raw text. Finally, keep in mind that, behind the scenes, LLMs often receive additional, invisible prompts or system messages alongside your visible query, which also shape the model's output.

Our goal is – by the end of this chapter – to train a basic LLM at least partially capable of similar text generation.

Some code in this chapter builds on:

- *The original OpenAI GPT-2 code: https://github.com/openai/gpt-2 (modified MIT License);*
- *PyTorch re-implementation of GPT-2 on Hugging Face: https://github.com/huggingface/transformers (Apache-2.0 license);*
- *PyTorch library source code for transformer components: https://pytorch.org/docs/stable (BSD-style license);*
- *Andrej Karpathy's GPT-2 re-implementations: https://github.com/karpathy/ng-video-lecture, https://github.com/karpathy/nanoGPT (MIT license).*

6.1 Text data

We will work with the text of "Alice in Wonderland" by Lewis Carroll (it is in public domain). Let us load it in.

```python
import requests # downloads
from bs4 import BeautifulSoup # html parsing
import re # regular expressions for text processing
import textwrap # wrap print output

# run the following code to download the text
# or you can read in the provided version directly

url = 'https://www.gutenberg.org/files/11/11-h/11-h.htm'
response = requests.get(url) # get web site's html
response.encoding = 'utf-8' # correct encoding
html_content = response.text # extract html content
soup = BeautifulSoup(html_content,'html.parser') # parse html
text = soup.get_text() # raw text from html as a string

# remove repeated sequential spaces and newline characters
text = re.sub(' +', ' ', text)
text = re.sub(r'\n\s*\n', '\n\n', text)

# save the file
with open("alice.txt", "w", encoding="utf-8") as file:
    file.write(text)

# read the saved file
# with open("alice.txt", "r", encoding="utf-8") as file:
#     text = file.read()

print("Text string length:", len(text))
print("Text:",textwrap.fill(text[:600], 58))
```

```
Text string length: 146458
Text:   Alice' s Adventures in Wonderland | Project Gutenberg
```

```
*** START OF THE PROJECT GUTENBERG EBOOK 11 ***  Alice's
Adventures in Wonderland by Lewis Carroll THE MILLENNIUM
FULCRUM EDITION 3.0  Contents   CHAPTER I.Down the Rabbit-
Hole   CHAPTER II.The Pool of Tears   CHAPTER III.A
Caucus-Race and a Long Tale   CHAPTER IV.The Rabbit Sends
in a Little Bill   CHAPTER V.Advice from a Caterpillar
CHAPTER VI.Pig and Pepper   CHAPTER VII.A Mad Tea-Party
CHAPTER VIII.The Queen's Croquet-Ground   CHAPTER IX.The
Mock Turtle's Story   CHAPTER X.The Lobster Quadrille
CHAPTER XI.Who Stole the Tarts?
```

In text modeling, one needs first to define the discrete units of analysis –
called **tokens** – that the text will be split into and that we will be predicting.
For instance, we could split the text into:

- Individual characters;
- Bigrams (character pairs);
- Trigrams (character sequences of length three);
- Full words;
- Syllables;
- Algorithmically defined subsets of characters (e.g., obtained using byte
  pair encoding (BPE) [SHB15]);
- And so on.

In this chapter, we will focus on *character-level* analysis and will be predicting
text one character at a time.

After determining the type of token we are working with, we need to define
the universe of possible tokens – in our case, characters – that appear in
text. This universe of possible tokens is called a *vocabulary*. We can also
call these tokens *items* of the vocabulary.

```python
vocab = sorted(list(set(text)))
V = len(vocab) # vocabulary size
print("Vocabulary:\n", textwrap.fill(str(vocab), 60))
print("Vocabulary size:", V)
```

```
Vocabulary:
```

```
['\n', '\r', ' ', '!', '(', ')', '*', ',', '-', '.', '0',
'1', '3', ':', ';', '?', 'A', 'B', 'C', 'D', 'E', 'F', 'G',
'H', 'I', 'J', 'K', 'L', 'M', 'N', 'O', 'P', 'Q', 'R', 'S',
'T', 'U', 'V', 'W', 'X', 'Y', 'Z', '[', ']', 'a', 'b', 'c',
'd', 'e', 'f', 'g', 'h', 'i', 'j', 'k', 'l', 'm', 'n', 'o',
'p', 'q', 'r', 's', 't', 'u', 'v', 'w', 'x', 'y', 'z', '|',
'\xa0', 'ù', '—', ' ' ', ' ' ', ' " ', ' " ']
Vocabulary size: 78
```

Notice that the uppercase and lowercase letters are distinct elements of the vocabulary – together with numbers, punctuation marks, and a space. Some characters you may not be familiar with include:

- Unicode character \xa0 is a non-breaking space   , which prevents line breaks at its position.
- \n is a newline character.
- \r is a carriage return used, e.g., in Windows newline convention.

We will associate with each token in the vocabulary a numerical *integer index* and helper functions to go between tokens and indices.

```
# token - integer index mapping
t2i = {tok:i for i,tok in enumerate(vocab)}
i2t = {i:tok for i,tok in enumerate(vocab)}

# functions to convert between integers and strings
str2ind = lambda s: [t2i[t] for t in s]
ind2str = lambda l: ''.join([i2t[i] for i in l])
```

Now we can encode and decode a string into and from integer space.

```
print(textwrap.fill(text[1500:1600], 60))
```

```
she ought to have wondered at this, but at the time it all
seemed  quite natural); but when the Rabb
```

```
print("Index encoding of a string:\n",
    textwrap.fill(str( str2ind(text[1500:1600]) ), 60))
```

```
Index encoding of a string:
 [62, 51, 48, 2, 58, 64, 50, 51, 63, 2, 63, 58, 2, 51, 44,
65, 48, 2, 66, 58, 57, 47, 48, 61, 48, 47, 2, 44, 63, 2, 63,
51, 52, 62, 7, 2, 45, 64, 63, 2, 44, 63, 2, 63, 51, 48, 2,
63, 52, 56, 48, 2, 52, 63, 2, 44, 55, 55, 2, 62, 48, 48, 56,
48, 47, 1, 0, 60, 64, 52, 63, 48, 2, 57, 44, 63, 64, 61, 44,
55, 5, 14, 2, 45, 64, 63, 2, 66, 51, 48, 57, 2, 63, 51, 48,
2, 33, 44, 45, 45]
```

```
# integers back to string
print( textwrap.fill(ind2str(str2ind(text[1500:1600])), 60))
```

```
she ought to have wondered at this, but at the time it all
seemed  quite natural); but when the Rabb
```

6.2 Sidebar: Byte pair encoding (BPE) tokenization

In practice, working at character level does not yield very powerful text prediction models, so the vocabulary is typically expanded to include not only individual characters, but also a variety of commonly encountered character combinations, which can be sub-word parts or even whole common words (e.g., "the"). This results in an expanded vocabulary. For example, OpenAI GPT-2 models use the vocabulary size of $V = 50257$ of distinct tokens [Rad+19]; Llama 3 models use the vocabulary size $V = 128K$ (K stands for thousands) [Gra+24].

Such expanded vocabularies are constructed algorithmically. **Byte pair encoding (BPE)** [SHB15] is one simple but powerful tokenization method used to split text into subword units instead of just words or characters. A version of BPE was used, for example, for training OpenAI's LLMs [Rad+19]. Implementation details can differ a lot, but below is a crux of the how the method works:

Table 6.1: Byte pair encoding (BPE) illustration

Step	Token sequence	Vocabulary of tokens
1	a b r a c a d a b r a _ r a	{_, a, b, c, d, r }
2	a b rac a d a b ra _ ra	{_, a, b, c, d, r, ra}
3	ab ra c a d ab ra _ ra	{_, a, b, c, d, r, ra, ab}
4	abra c a d abra _ ra	{_, a, b, c, d, r, ra, ab, abra}
5	abrac a d abra _ ra	{_, a, b, c, d, r, ra, ab, abra, abrac}

1. *Initialization.* Begin with the full text as a single string decomposed into individual characters. Unique characters are treated as distinct tokens and are included in the initial vocabulary.

2. *Iterative merging.* Identify the most frequent pair of adjacent tokens and merge them into a new token, which (a) is added to the vocabulary and (b) replaces the corresponding token pairs in the text. In case of several equally frequent most frequent pairs, choice is arbitrary – for example, we could agree to merge the left-most most frequent pair first. Repeat until the desired vocabulary size V is reached.

Table 6.1 illustrates the BPE in action using a toy example `abracadabra ra` string, representing the space via an underscore, and merging until vocabulary size $V = 10$ is reached. In this example, the full alphabet only has 6 basic characters: `a`, `b`, `c`, `d`, `r`, `_`.

After the iterative merging and expansion of the vocabulary in Table 6.1, `abra` appears as its own token in the vocabulary. Now, if a sequence `abra` is encountered, it can be encoded as its own token; but if some variation of this word is present, e.g., due to a typo, such as `abbra`, we still could encode it as three tokens `ab`, `b`, `ra`. As a result, the BPE can efficiently handle during text encoding both (a) rare words that don't have dedicated tokens (by breaking them into familiar pieces) and (b) common words (by creating dedicated tokens for them).

Here is a simple Python implementation of BPE:

```python
from collections import Counter
```

```python
# frequency of adjacent token pairs
def get_freq(sequence):
    pairs = Counter()
    for i in range(len(sequence) - 1):
        pair = (sequence[i], sequence[i + 1])
        pairs[pair] += 1
    return pairs

# merge the most frequent token pair
def merge_tokens(sequence, pair_to_merge):
    a, b = pair_to_merge
    merged_token = a + b
    i = 0
    new_sequence = []
    while i < len(sequence):
        if i < len(sequence) - 1 and sequence[i] == a and
        ↪    sequence[i + 1] == b:
            new_sequence.append(merged_token)
            i += 2
        else:
            new_sequence.append(sequence[i])
            i += 1
    return new_sequence

# example
sequence = "a b r a c a d a b r a _ r a".split()
vocab_bpe = sorted(set(sequence))
V_bpe = 10   # target vocabulary size
merge_rules = [] # to store merge history

print(f"Step 1: {' '.join(sequence)}")
print(f"Vocabulary: {sorted(vocab_bpe)}\n")

step = 2
while len(vocab_bpe) < V_bpe:
    stats = get_freq(sequence)
    if not stats:
```

```
        break
    most_common = stats.most_common(1)[0][0]
    merge_rules.append(most_common)
    sequence = merge_tokens(sequence, most_common)
    vocab_bpe.append(most_common[0] + most_common[1])
    print(f"Step {step}: {' '.join(sequence)}")
    print(textwrap.fill(f"Vocabulary: {vocab_bpe}",60)+"\n")
    step += 1
```

```
Step 1: a b r a c a d a b r a _ r a
Vocabulary: ['_', 'a', 'b', 'c', 'd', 'r']

Step 2: a b ra c a d a b ra _ ra
Vocabulary: ['_', 'a', 'b', 'c', 'd', 'r', 'ra']

Step 3: ab ra c a d ab ra _ ra
Vocabulary: ['_', 'a', 'b', 'c', 'd', 'r', 'ra', 'ab']

Step 4: abra c a d abra _ ra
Vocabulary: ['_', 'a', 'b', 'c', 'd', 'r', 'ra', 'ab',
'abra']

Step 5: abrac a d abra _ ra
Vocabulary: ['_', 'a', 'b', 'c', 'd', 'r', 'ra', 'ab',
'abra', 'abrac']
```

Text may be further pre-processed prior to running BPE, for example, by incorporating special tokens to mark the ends of sentences, documents, etc. Such tokens are useful in marking points, where text prediction / generation should be ended. Numbers may be handled differently – for example, each digit may be assigned a token, but number pairs may not be allowed to merge [Tou+23b].

Upon the BPE application to the body of text and the construction of the vocabulary of desired size *V*, a deterministic sequence of merges is established / learned. To tokenize any new encountered text, we apply to such text that same sequence of merges.

```python
# apply learned merge rules to a new sequence
def apply_merges(sequence, merge_rules):
    for a, b in merge_rules:
        i = 0
        new_sequence = []
        while i < len(sequence):
            if i < len(sequence) - 1 and sequence[i] == a and
            ↪  sequence[i + 1] == b:
                new_sequence.append(a + b)
                i += 2
            else:
                new_sequence.append(sequence[i])
                i += 1
        sequence = new_sequence  # update after each rule
    return sequence

print(textwrap.fill("Learned merge rules: "+
    str(merge_rules),60))

# tokenize new sequence
new_seq = "a b b r a _ r a".split()
tokenized = apply_merges(new_seq, merge_rules)

print("New text:", ' '.join(new_seq))
print("Tokenized:", tokenized)
```

```
Learned merge rules: [('r', 'a'), ('a', 'b'), ('ab', 'ra'),
('abra', 'c')]
New text: a b b r a _ r a
Tokenized: ['ab', 'b', 'ra', '_', 'ra']
```

To accommodate the wide variety of scripts used around the world, BPE is often applied not directly to characters, but to the sequence of byte values obtained from UTF-8 encoding of the text. (Recall that UTF-8 encodes every character as sequence of 1 to 4 bytes.) Since a byte can represent 256 distinct values ($2^8 = 256$), the initial vocabulary consists of just 256 tokens – one for each possible byte value. Such compact vocabulary is immediately sufficient to represent any UTF-8-encodable text, which includes nearly all text encountered in practice.

This byte-level initialization is far more compact than, say, assigning a separate token to each character in a complex script. For example, encoding every Chinese logograph individually would require thousands of tokens – around 3,000 characters are commonly used, and Unicode includes nearly 100,000 Han characters in total. Representing each of them individually would require an unmanageably large vocabulary.

The byte-level approach is particularly useful in real-world scenarios, where tokenization is applied not to sanitized literary texts like *Alice in Wonderland*, but to massive, messy data sets collected from the open web – often containing diverse characters and symbols. Here is a byte encoding example:

```
# example: "Don't panic" (biè huāng) in Chinese
chi = "别慌"

utf8_bytes = chi.encode('utf-8')
print("UTF-8 bytes:", utf8_bytes)

# convert each byte into an integer token (0-255)
seq = list(utf8_bytes)
print("Byte-level tokens:", seq)

# decode back (for illustration)
decoded = bytes(seq).decode('utf-8')
print("Decoded string:", decoded)
```

```
UTF-8 bytes: b'\xe5\x88\xab\xe6\x85\x8c'
Byte-level tokens: [229, 136, 171, 230, 133, 140]
Decoded string: 别慌
```

There is also a lot of ongoing research into improving on the BPE tokenizer, which works well and is efficient in practice, but is also ultimately just a heuristic, ad hoc procedure, which is not explicitly optimized for a given objective function in text modeling [Cla+22; Tay+21; Xue+22; Kud18].

For those curious, OpenAI offers an online tool to observe their BPE-based tokenizers in action: https://platform.openai.com/tokenizer. Their `tiktoken` library provides Python implementation of OpenAI's tokenization algorithms. Alternatively, Python `sentencepiece` library by Google also offers implementation of BPE and as well as other sophisticated tokenization algorithms. Hugging Face offers a nice BPE tutorial: https://huggingface.co/learn/llm-course/en/chapter6/5.

Overall, depending on which tokenization scheme we use, the vocabulary set will differ, and strings will get encoded into distinct integer sequences (e.g., based on index over a larger set of varied-size word pieces rather than over single characters alone). In the end, however, the encoded text data still consists of integer sequences, and the modeling techniques that we discuss next can be applied to such sequences regardless of which tokenization scheme is used to encode the text. For simplicity, we will stick to the character-level analysis for our running example of the *Alice in Wonderland* text.

6.3 Text prediction as multinomial logistic regression

What we want to do now is build a function that takes in some text and puts out more text. Given that we are working at the character token level, we want a function that takes in as input a sequence of characters and puts out more characters. Such a function is going to be our text prediction model.

6.3.1 Token sampling procedure

Specifically, we will define our text prediction model as a function that outputs a discrete probability distribution over all possible next tokens (in our case, characters) in the vocabulary. We can use this probability distribution to sample the next character based on an input sequence of characters. During training, the distribution output by this function would be fit / calibrated to the data – so it reflects the observed relative frequencies of different character sequences. (Such use of unlabeled text data to generate the training signal for the model is called *self-supervised learning* [Wike].)

Here is an example of the text generation procedure that uses a distribution over next characters:

- Start with an input sequence, e.g., `appl` .

  - A well-calibrated probability distribution, conditioned on this input, should hopefully assign larger next-letter probabilities to the letters like `y` , `e` , or `i` that we would not be surprised by, compared to the more surprising letters like `d` or `x` .

- Sample a new letter from the distribution, conditioned on `appl` . Say, we get `i` . Then, the new string is `appli` .

- Repeat the process again and sample another next character, this time conditioning on the input sequence `appli` . Perhaps, we sample `e` and get `applie` .

- Repeat the sampling procedure one more time, conditioning on `applie` input. Perhaps we sample `d` , getting a string `applied` .

This is really all there is to the text generation procedure we want to implement – it is also the basis of how the current state-of-the-art (SOTA) large language models (LLMs) operate.

---

i Going beyond the next-token prediction paradigm

The predominant next-token prediction paradigm is known as *autoregressive* modeling, borrowing the terminology from the statistical time-series literature. There is active research on going beyond token-by-token generation – examples include *speculative decoding* techniques [LKM23] and *diffusion models* [Nie+25; Lab+25].

6.3.2 Predicting next-token probabilities with softmax

If we could somehow represent text in a numerically convenient way to input into a neural net, then the rest of the task is simple in principle – a neural net could take as input the numerical representation of the current text sequence and output a discrete probability distribution over the tokens in the vocabulary. The idea of using neural nets for such character-level text prediction goes back at least to [SH96].

For example, the neural net could output such a discrete distribution over tokens using the softmax function:

$$P(\text{next token} = j \mid \text{current token sequence}) = \frac{\exp(u_j/\tau)}{\sum_{k=1}^{V} \exp(u_k/\tau)},$$

where u_j is a numerical score assigned to the token j by the neural net and $\tau > 0$ (pronounced *tau*) is a temperature hyperparameter (larger τ values make all tokens more equally likely; for now, we set $\tau = 1$). We originally encountered such softmax functional form in the context of the multinomial logistic regression. We could use this distribution to sample the next character from the vocabulary.

6.3.3 Representing tokens numerically using embeddings

Technically, we already have a numerical representation of characters – their integer indices in the vocabulary. However, inputting sequences of these integers directly as numbers for multiplication and addition by the neural net weights does not work well.

Here is a better idea – associate with each token (in our case, with each character) a randomly initialized vector – a list of numbers – of some size E; this vector is called an **embedding**. Treat this vector as a numerical representation of the character and optimize the vector's entries as parameters such that they help the neural net predict the next character in a body of existing text as accurately as possible. Brilliant!

We can indeed take derivatives not only with respect to parameters of the neural net itself, but also with respect to inputs – so we can update vectors representing characters to minimize the prediction error.

Altogether, for a vocabulary of size V, we will have V vectors of length E – each E-long vector will, upon training, capture some statistical essence of its associated character that helps us predict new characters that are likely to follow it in the real text. An embedding vector can be thought of as coordinates of the corresponding token in the E-dimensional space. The array of embeddings for our vocabulary is then just a $V \times E$ floating point matrix that we need to optimize.

Here is an illustration of randomly initialized embedding vectors (to be fine-tuned) for a string `hello` , with embedding size (vector length) $E = 6$:

$$
\begin{matrix}
\mathbf{h} & \mathbf{e} & \mathbf{l} & \mathbf{l} & \mathbf{o} \\
\begin{bmatrix} 0.12 \\ -0.45 \\ 0.33 \\ 0.91 \\ -0.17 \\ 0.08 \end{bmatrix} &
\begin{bmatrix} -0.22 \\ 0.14 \\ 0.78 \\ -0.66 \\ 0.05 \\ -0.30 \end{bmatrix} &
\begin{bmatrix} 0.41 \\ 0.09 \\ -0.13 \\ 0.37 \\ -0.50 \\ 0.29 \end{bmatrix} &
\begin{bmatrix} 0.41 \\ 0.09 \\ -0.13 \\ 0.37 \\ -0.50 \\ 0.29 \end{bmatrix} &
\begin{bmatrix} -0.31 \\ 0.67 \\ 0.22 \\ -0.44 \\ 0.19 \\ -0.11 \end{bmatrix}
\end{matrix}
$$

Each number in these vectors is a trainable parameter. Notice that vectors for `l` are the same – because, in a vocabulary, each item is associated with its own single, distinct embedding vector.

In general, a larger embedding size E tends to offer greater expressive power, enabling the model to capture more nuanced relationships between tokens. However, this benefit must be weighed against the amount and quality of training data, as well as the risk of overfitting. In low-data settings, smaller embedding sizes are often more appropriate, and increasing the embedding dimensionality (vector length) may not improve – and can even degrade – prediction performance on unseen data. In practice, small-scale models often use embeddings with a few hundred dimensions, while commercial-grade language models commonly employ embeddings with thousands of dimensions. For example:

- OpenAI's largest GPT-3 model `GPT-3 175B` has an embedding size of $E = 12288$, while the smallest `GPT-3 Small` has $E = 768$ – see Table 2.1 in [Bro+20].
- Llama 3 405B parameter model by Meta has an embedding size of $E = 16384$ [Gra+24].

The idea of learnable embeddings for text prediction, in the form of word feature vectors, was proposed in [Ben+03], although the idea of continuous vector representations for discrete entities goes much further back in time (the provided reference contains further pointers). Other early papers that demonstrated the power of and popularized embeddings include [Mik13;

PSM14]. The embeddings were then incorporated in the modern transformer architecture for LLMs, which we will study later, and the rest is history [Vas+17; Dev+19; Rad+19].

6.3.4 An outline of the computation

Next, we will implement the idea of representing characters using embedding vectors and construct a simple model in the style of the multinomial logistic regression to predict the next character from preceding characters in text. The computation encompasses the following steps:

Step 1. Sample raw strings. Sample a batch of B strings, each string starting at a random point in text, each string of length S characters. These B strings of S characters will be stored as an array X_{raw} of shape $B \times S$ (batch size $\times$ sequence length). For example, consider a hypothetical sample of 4 sequences $B = 4$, of $S = 6$ characters each, forming X_{raw} array:

```python
import numpy as np

X_raw = np.array([
["H", "e", "l", "l", "o", "."],
["i", "d", " ", "i", "t", "\n"],
["!", " ", "W", "h", "y", "?"],
["a", "r", "e", " ", "y", "o"],
])
print(X_raw)
print(X_raw.shape)
```

```
[['H' 'e' 'l' 'l' 'o' '.']
 ['i' 'd' ' ' 'i' 't' '\n']
 ['!' ' ' 'W' 'h' 'y' '?']
 ['a' 'r' 'e' ' ' 'y' 'o']]
(4, 6)
```

Step 2. Raw strings to integers. Replace characters with their corresponding indices – numerical positions in the vocabulary – to get an integer array X_{ind}, also shaped $B \times S$:

```
B, S = X_raw.shape
X_ind = np.zeros((B,S), dtype=int)
for b in range(B):
    for s in range(S):
        X_ind[b,s] = t2i[X_raw[b,s]] # token to index

print(X_ind)


[[23 48 55 55 58  9]
 [52 47  2 52 63  0]
 [ 3  2 38 51 68 15]
 [44 61 48  2 68 58]]
```

Here each number is a vocabulary index for a character – for example, capital H is index 23 in the vocabulary, which can be looked up in our dictionary via t2i['H'] .

Step 3. Character integers to embedding vectors. For each integer index for a character in X_{ind}, look up its corresponding embedding vector of size E from the overall $V \times E$ vocabulary array of embeddings. Replace this character scalar index with its corresponding embedding vector of length E. This means that each cell of this 2D array of character sequences now contains a vector – this can be represented by adding a third dimension going depth-wise and can be written as a 3D array X_{emb} shaped $B \times S \times E$ (batch size $\times$ sequence length $\times$ embedding vector length). Now, the array will contain – instead of integer indices for characters – the characters' vector representations. For example, for $B = 4$, $S = 6$, $E = 2$, X_{emb} can be understood as four $S \times E$ matrices stacked one upon another:

```
np.random.seed(999)

E = 2 # embedding size

# array of all embeddings shaped as
# Alice in Wonderland vocab size V x embedding size E
token_embeddings = np.random.normal(size=(V, E))
```

```
# array of token indices to array of embeddings
X_emb = token_embeddings[X_ind]
print(X_emb.shape)
print(X_emb)
```


```
(4, 6, 2)
[[[ 0.60487573  0.30054313]
  [ 0.10892936 -0.03912716]
  [ 0.43816732  0.18814676]
  [ 0.43816732  0.18814676]
  [ 0.48026069 -1.08525169]
  [ 0.98434258 -0.83544737]]

 [[-0.92767682 -0.71360919]
  [ 0.61275229  0.25506099]
  [-0.26613444 -0.64890071]
  [-0.92767682 -0.71360919]
  [ 0.66565147  1.07865774]
  [ 0.12715784  1.40189088]]

 [[ 1.56626757 -2.09137019]
  [-0.26613444 -0.64890071]
  [ 0.15188388  1.08328312]
  [ 0.07300425  0.73492972]
  [-0.30719913  0.52414483]
  [ 0.55931017  0.4740131 ]]

 [[ 2.0691885  -0.30205227]
  [ 0.04509789  0.8590863 ]
  [ 0.10892936 -0.03912716]
  [-0.26613444 -0.64890071]
  [-0.30719913  0.52414483]
  [ 0.48026069 -1.08525169]]]
```

Here, each of the four blocks represents a sequence of six characters, each character represented by an embedding vector of length two. For example, newline character is the first element in the vocabulary, and so its embedding is given by:

```
token_embeddings[0]
```

```
array([0.12715784, 1.40189088])
```

You can confirm the position of this newline character embedding in the generated X_{emb} array above – at the end of the second sequence in the batch.

Figure 6.1 visualizes X_{emb} array structure in 3D – each little cubic cell is a floating point scalar in X_{emb} array.

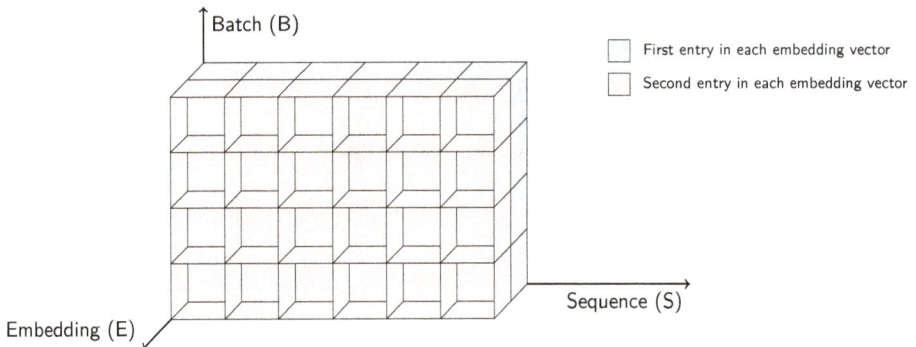

Figure 6.1: 3D batch array visualization

Step 4. Average over sequences. For each sequence in X_{emb}, compute the average embedding of its characters:

$$X_{\text{avg.emb}} = \text{average}(X_{\text{emb}}, \text{dimension} = \text{sequence}).$$

This collapses the 3D array $X_{\text{emb}}(B \times S \times E)$ into a 2D array $X_{\text{avg.emb}}(B \times E)$:

```
# average embedding within each sequence
X_avg_emb = np.mean(X_emb, axis=1)

print(X_avg_emb.shape)
print(X_avg_emb)
```

```
(4, 2)
[[ 0.50912383 -0.2138316 ]
 [-0.11932108  0.10991509]
 [ 0.29618872  0.01268331]
 [ 0.35502381 -0.11535012]]
```

Step 5. Linear layer of neurons. Pass $X_{\text{avg.emb}}$ array $(B \times E)$ containing an average embedding per input sequence to a linear layer of neurons with parameters $W_{E \times V}$ and $b_{1 \times V}$, to get

$$\underset{B \times V}{X_{\text{logits}}} = \underset{B \times E}{X_{\text{avg.emb}}} \cdot W_{E \times V} + b_{1 \times V}.$$

The layer treats E entries in each average embedding vector as variables and outputs V neurons – one for each possible next character in the vocabulary. The layer thus outputs an array X_{logits} shaped $B \times V$ of utility scores (also called logits) that can assume values in $(-\infty, \infty)$ range and are proportional to the relative likelihood of different characters to follow the current input text:

```
np.random.seed(123)

# layer params
W = np.random.normal(size=(E, V))
b = np.random.normal(size=(1, V))
# note that bias is commonly initialized to zero

# BxV array of logits / utilities
X_logits = X_avg_emb.dot(W) + b

print(X_logits.shape)
```

```
(4, 78)
```

Figure 6.2 visualizes this neural net layer:

Step 6. Utilities to probabilities. Apply softmax function

$$P(j) = \frac{\exp(x_j)}{\sum_{k=1}^{V} \exp(x_k)}$$

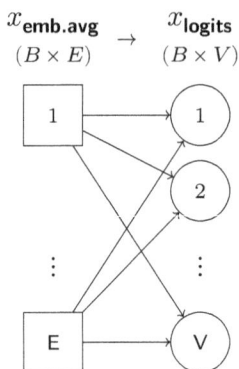

Figure 6.2: Multinomial logistic regression as a single-layer neural net

to elements of each row of $B \times V$ array X_{logits} of utilities to convert it to array of probabilities X_{prob} – which indicates probabilities of the next character for each sequence in the batch. Specifically, this means taking exponent elementwise of $B \times V$ array X_{logits} and normalizing (dividing) its rows by row-sums so row elements add up to one. X_{prob} can be used to sample next characters of sequences in the batch. (This is also exactly the output of a multinomial logistic regression.)

$$X_{\text{prob}} = \text{softmax}(X_{\text{logits}}, \text{dimension} = \text{vocab}).$$
$$\underset{B \times V}{} \qquad \underset{B \times V}{}$$

```
def row_wise_softmax(x):
    # subtract the maximum value for numerical stability
    # this operation does not affect the resulting
    ↪  probability value
    # because it is equivalent to dividing numerator and
    ↪  denominator
    # by the same exp(max) value
    e_x = np.exp(x - np.max(x, axis=1, keepdims=True))
    # divide by the sum of exponentials for normalization
    return e_x / np.sum(e_x, axis=1, keepdims=True)

X_prob = row_wise_softmax(X_logits)

print(X_prob.shape)
```

```
# to verify probs. sum to 1 across full vocab.
print(X_prob.sum(1))
```

```
(4, 78)
[1. 1. 1. 1.]
```

We can also sample from these probabilities – either using a built-in choice function:

```
samples = np.array([
    np.random.choice(X_prob.shape[1], p=row) for row in
 ↳ X_prob
])

print(samples)  # e.g., array([1, 2, 0])
```

```
[61  0 10 55]
```

or manually via uniform random draws and the cumulative distribution function:

```
# row-wise cumulative probabilities
# shape: (batch size B, vocab size V)
cumulative = np.cumsum(X_prob, axis=1)

# one uniform random number per row
u = np.random.rand(B)

# compare uniform draw with cumulative distribution to sample
samples = (cumulative < u[:, None]).sum(axis=1)

print(samples)
```

```
[ 2 50 75 61]
```

Step 7. Error function. Compute the cross-entropy loss function $CE(y_{ind}, X_{prob})$ comparing predicted probabilities to the actual observed next characters y_{ind} (encoded as indices, shaped $B \times 1$) that follow sequences in the input array X_{ind} in the training text. Cross entropy is just the negative log probability of the observed next character for a sequence. Let o be the vocabulary index of the observed next token for a sequence, given by the corresponding entry in y_{ind}. Then CE for that sequence is:

$$-\log P(o) = -\log\left[\frac{\exp(x_o)}{\sum_{k=1}^{V}\exp(x_k)}\right] = -x_o + \log\left[\sum_{k=1}^{V}\exp(x_k)\right],$$

where x values represent corresponding utilities from X_{logits} array. This quantity is computed for each sequence independently and is then averaged over the batch to get the final CE error estimate.

> **i Note**
>
> Cross-entropy can be computed on raw logits / utility scores X_{logits} rather than on the probability values X_{prob}. One needs to worry about the numerical stability in log-sum-exp expression though. It is usually better to use libraries' smart implementations of cross-entropy. Always check if the provided cross-entropy function expects logits or probabilities; logits are more commonly required.

Cross-entropy computation in code:

```
# continuation of sequences in X_raw
y_raw = np.array([
[" "],
["f"],
["\n"],
["u"]
])


# indices
y_ind = np.zeros((B,1), dtype=int)
for b in range(B):
    y_ind[b,0] = t2i[y_raw[b,0]]
```

```
# cross entropy for each batch sequence
# i.e., negative log probability assigned by the model
# to the actual realized next character
CE_s = -np.log(X_prob[np.arange(B), y_ind.flatten()])
print(CE_s)

# average cross entropy
CE = np.mean(CE_s)
print(CE)
```

```
[3.61284783 5.0697951  4.55075799 5.66530036]
4.724675319705576
```

Step 8. Parameter update. The final step is to update both the vocabulary embeddings and the neural net parameters (here, parameters of the single fully connected linear layer) to minimize the cross-entropy loss for sampled batches of string sequences. This requires the computation of derivatives. Instead of continuing with the toy example in NumPy, we will now move to PyTorch and implement the training for the presented simple text prediction model from the average within-sequence embeddings on the *Alice in Wonderland* text.

6.3.5 PyTorch implementation

To implement this setup, we will use PyTorch library in Python. We will select the fastest available device on the machine to run the neural net – in general, GPU is faster than MPS (Apple-specific), which is faster than CPU.

```python
import torch
import torch.nn as nn
from torch.nn import functional as F

if torch.cuda.is_available():
    device = 'cuda' # gpu
```

```
elif torch.backends.mps.is_available():
    device = 'mps'
else:
    device = 'cpu'

print("Fastest available hardware for deep learning:",
 ↳  device)
```

Fastest available hardware for deep learning: mps

Train-test split of the data – for simplicity, first 80% of text is the train set, and the remaining 20% is the test set:

```
# integer value pytorch tensor
# containing integer encoding of the full
# Alice in Wonderland text
data = torch.tensor(str2ind(text), dtype=torch.long)
n = int(0.8*data.shape[0])
data_train = data[:n]
data_test = data[n:]

print("First 5 character indices:", data_train[:5])
```

First 5 character indices: tensor([0, 0, 16, 55, 52])

Let us write a function to sample a batch of text sequences to train on:

```
B = 4 # batch size - number of token sequences
S = 6 # input sequence length

def get_batch(split):
    # get text sequences from random starting points
    if split == 'train':
        d = data_train
    if split == "test":
```

```
          d = data_test

     # sampling starting points for sequences
     start_ind = torch.randint(d.shape[0] - S, (B,))
     # (d.shape[0] - S) is 1 above the largest integer to be
     ↪  drawn

     X_ind = torch.stack([d[i:i+S] for i in start_ind])
     y_ind = torch.stack([d[i+S] for i in start_ind])
     X_ind, y_ind = X_ind.to(device), y_ind.to(device)
     return X_ind, y_ind

get_batch('train')
```

```
(tensor([[62, 64, 61, 59, 61, 52],
         [ 2, 76, 27, 48, 63,  2],
         [48,  2, 53, 64, 56, 59],
         [58, 61,  2, 63, 51, 48]], device='mps:0'),
 tensor([62, 64, 52,  1], device='mps:0'))
```

This gives us a 4×6 array X_{ind} that contains 4 random sequences, each of length 6, starting from random points in the text – each sequence is represented by a list of integers that correspond to characters. And we also get a length 4 vector y_{ind}, which is our prediction target and contains the index for the character that follows each of the 4 input sequences.

We will now define a neural net in PyTorch that will contain an array of embeddings for the vocabulary (randomly initialized) and a single output layer of neurons processing input variables. Each neuron will output the logit / utility score proportional to likelihood of the character the neuron represents. These scores across neurons can be converted to probabilities using the softmax function $P(j) = \exp(x_j) / \sum_{k=1}^{V} \exp(x_k)$.

Each sequence input to the neural net will constitute an observation and the average embedding across sequence characters will constitute the variables describing that observation. So, for example, for a sequence of length 6 and an embedding vector of length 16, we will take the average of 6 16-dimensional vectors, obtaining the 16-dimensional average vector – its 16 entries are 16 variables representing the sequence.

The input sequence batch, in integer index representation, will have 2D shape (batch size B × sequence length S). However, when we replace each character integer with a corresponding embedding vector, the input sequence batch will become a 3D array of shape (batch size B × sequence length S × embedding size E). Taking an average will reduce the array along the second dimension, so we will get (batch size B × embedding size E) array of average within-sequence embeddings, which we will use as the input to get the prediction.

Specifically, an average embedding per sequence will be input to a simple linear layer of neurons in order to compute numerical scores for all items in the vocabulary (size V) – the scores can be mapped to probabilities using softmax. The linear layer here simply performs linear-regression-style dot product combined with addition of the bias terms:

$$X_{\text{logits}} = X_{\text{emb.avg}} \cdot W_{E \times V} + b_{1 \times V},$$
$$_{B \times V} _{B \times E}$$

where $X_{\text{emb.avg}}$ is the array where each row holds an average embedding for the corresponding sequence.

Let us see it in code:

```
# random seed
torch.manual_seed(999)

# V is Alice in Wonderland vocabulary size
B = 4  # batch size - number of token sequences
S = 8  # input sequence length
E = 16 # embedding size

# model definition
class AvgEmbeddingModel(nn.Module):

    def __init__(self, V, E):
        super().__init__()
```

```
        # this calls the __init__ of the parent class
        ↪ nn.Module

        # V x E array of all embedding vectors
        self.token_embeddings = nn.Embedding(V, E)

        # a fully-connected layer of V neurons
        # each neuron takes E input values
        self.output_layer = nn.Linear(E, V)

    def forward(self, X_ind):
        # X_ind is B x S array
        # sequences of integers

        # X_emb is B x S x E array
        # sequences of embeddings
        X_emb = self.token_embeddings(X_ind)

        # X_emb_avg is B x E array
        # average within-sequence embeddings
        X_emb_avg = torch.mean(X_emb, 1)

        # X_logits is B x V array
        # utilities / logits for possible next characters
        # row-wise softmax would generate probabilities
        X_logits = self.output_layer(X_emb_avg)
        return X_logits

    def next_token_prob(self, X_logits):
        # X_prob is B x V array of probabilities
        # for possible next characters
        X_prob = F.softmax(X_logits, dim=-1) # B x V
        return X_prob .

    def loss(self, X_logits, y_ind):
        # cross-entropy loss
        return F.cross_entropy(X_logits, y_ind)
```

```
net = AvgEmbeddingModel(V, E)
net = net.to(device)
```

In the definition above, `__init__` function initializes the neural net and specifies all its parameters; `forward` function specifies the forward computation, returning logits; `next_token_prob` function compute next-token probabilities based on logits; `loss` function computes the cross-entropy loss via PyTorch implementation, which accepts the prediction logits and the index for the realized tokens.

```
print('Number of trainable parameters (embeddings + output
 ↪  layer)\n in AvgEmbeddingModel:',
   sum(p.numel() for p in net.parameters()))
```

```
Number of trainable parameters (embeddings + output layer)
 in AvgEmbeddingModel: 2574
```

Now, to training:

```
import time

# training settings
learning_rate = 1e-2 # 0.01
max_iters = 3000
eval_interval = 300
eval_iters = 200

# AdamW optimizer
optimizer = torch.optim.AdamW(net.parameters(),
 ↪  lr=learning_rate)

# train and test set evaluation
@torch.no_grad() # this decorator disables gradient
 ↪  calculation within the function
```

```python
def evaluate(net):
    record = {}
    net.eval() # evaluation mode (e.g., turns off dropout
 ↪  when present)
    for split in ['train', 'test']:
        losses = torch.zeros(eval_iters)
        for k in range(eval_iters):
            X_ind, Y_ind = get_batch(split)
            X_logits = net(X_ind)
            loss = net.loss(X_logits, Y_ind)
            losses[k] = loss.item()
        record[split] = losses.mean()
    net.train() # train mode
    return record

# training loop
start_time = time.time()
for i in range(max_iters):

    # regularly evaluate the loss on train and test sets
    if i % eval_interval == 0 or i == max_iters - 1:
        eval_ = evaluate(net)
        print(f"Step {i}: "
                f"Train loss = {eval_['train']:.4f}, "
                f"Test loss = {eval_['test']:.4f}")

    # sample a batch of data
    X_batch, y_batch = get_batch('train')

    # evaluate the loss and update the parameters
    X_logits = net(X_batch)
    loss = net.loss(X_logits, y_batch)
    optimizer.zero_grad()
    loss.backward()
    optimizer.step()

elapsed = time.time() - start_time
```

```
print(f"\n\t Elapsed time = {elapsed/60:.2f} mins")
```

```
Step 0: Train loss = 4.4215, Test loss = 4.3955
Step 300: Train loss = 3.0962, Test loss = 3.1353
Step 600: Train loss = 3.1422, Test loss = 3.1990
Step 900: Train loss = 3.0638, Test loss = 3.0082
Step 1200: Train loss = 2.9821, Test loss = 3.0287
Step 1500: Train loss = 2.8936, Test loss = 3.0107
Step 1800: Train loss = 3.0589, Test loss = 3.0315
Step 2100: Train loss = 2.9676, Test loss = 2.9623
Step 2400: Train loss = 2.9257, Test loss = 3.0203
Step 2700: Train loss = 2.8846, Test loss = 2.9998
Step 2999: Train loss = 2.9716, Test loss = 2.9562

	 Elapsed time = 0.32 mins
```

```
print("Final test set cross-entropy (CE):",
    float(f"{eval_['test']:.4f}"))
print("Probability of picking the correct next token: ")
print("> Neural net (exp(-CE)):",
    float(f"{torch.exp(-eval_['test']):.4f}"))
print("> Uniform random guess (1/V):", round(1/V,4))
```

```
Final test set cross-entropy (CE): 2.9562
Probability of picking the correct next token:
> Neural net (exp(-CE)): 0.052
> Uniform random guess (1/V): 0.0128
```

Consider the final test set cross-entropy (CE) for the model, which is the average negative log-probability assigned by the model to the correct token. Then, on average, the correct token is picked by the neural net with probability $P(\text{observed outcome}) \approx \exp(-\text{CE})$. In contrast, a uniform random guess would only identify the correct token with probability $1/V$. As you can see, the neural net is doing better than such a random guess.

Despite this encouraging news, if we try to generate some text this way, we will get gibberish.

```
def generate_text(net, n_new_tokens):
    X_ind = torch.zeros((1, 1), dtype=torch.long,
↪   device=device)
    for i in range(n_new_tokens):
        # crop to sequence size
        X_ind_crop = X_ind[:, -S:] # 1 x S
        X_logits = net(X_ind_crop) # 1 x V
        X_prob = net.next_token_prob(X_logits) # 1 x V
        X_ind_next = torch.multinomial(X_prob, num_samples=1)
↪   # 1 x 1
        X_ind = torch.cat((X_ind, X_ind_next), dim=1) # 1 x
↪   S+1
    return ind2str(X_ind[0].tolist())

net.eval()
s = generate_text(net, n_new_tokens=500)
print(textwrap.fill(s, 60))
```

```
Areie se npryataw  nidtha ,sgeetil We hens mof uoot aotn—ug
Wlhieatigs etb hhro seo natpd ?h tet olnaadtak nnatlheat" ! '
tohdahey rhws  ehWeh pesi  oano tu e mhhMsoeok sjn—dhee nbas
Ao e  sofhsol r entihoH use,laan  niatgmhlel r ouosa
irupnstwe ehrerud lo, hb etrrobe "f  hror mefelann role
oonttohi,: a wts tnae ihttdr Kli iivttns si aaatnins yohe
aAt tseireon w siomei nt eshd "a xertpab esitrfs eha out 'vnss
a g  i "ygo 'e'' oroattae peyntt oehse inthtoeng zin
TEeithir  nta  edohont ehpa rohtsi
```

What is going wrong? While the batch size B and the embedding size E could be increased, the biggest issue is that our predictive function is not expressive enough – the neural net is definitely too shallow, and, further, taking the average of embeddings across full sequences destroys the information about the order of tokens. Ideally, we want to build a model that is deep and that can automatically decide how to combine token embedding vectors for prediction, while also being aware of the token ordering. The celebrated *transformer* architecture, based on the *attention* mechanism, achieves just that, as we will see next.

Nevertheless, the work we have done is not futile – we have built a working text prediction model skeleton. Everything that comes next is just some modifications of this skeleton to get the neural net to do a better job.

6.4 Sidebar: Decorators

You may have noticed a construction `@torch.no_grad()` in the code earlier. This expression defines a *function decorator* – a function that takes another function and extends its behavior without explicitly modifying it. For example:

```
def my_decorator(func):
    def wrapper():
        print("Something happens before func runs")
        func()
        print("Something happens after func runs")
    return wrapper

@my_decorator
def say_hello():
    print("Hello!")

say_hello()
```

```
Something happens before func runs
Hello!
Something happens after func runs
```

`@my_decorator` is just syntactic sugar Python provides for writing `say_hello = my_decorator(say_hello)`. It wraps `say_hello()` in the `wrapper()` function, adding behavior before and after its call.

In PyTorch, `@torch.no_grad()` is a decorator that disables gradient calculation – it prevents PyTorch from tracking operations on tensors, so no computation graph is built, which reduces memory usage and speeds up computation. This is useful where gradients are not needed, for example, during model testing.

In Python, beyond the basic function decorators, there are also other types of decorators, such as class decorators, built-in decorators, parameterized decorators, etc. – but those are out of scope for us.

6.5 Transformer architecture for text prediction

We now proceed to implement a mini-version of a large language model (LLM) based on the transformer architecture. *Attention is all you need* paper [Vas+17] that introduced the transformer neural network architecture in 2017 has achieved, incredibly, over 187 thousand citations as of July 2025, underscoring the critical value of this innovation. This transformer architecture is behind the majority of the current state-of-the-art LLMs, such as OpenAI's GPT-4 models [Ach+23], Meta's Llama models [Gra+24], and DeepSeek-V3 [Liu+24]. In fact, not accidentally, "GPT" in the names of OpenAI's LLMs stands for Generative Pre-trained *Transformer* [Rad+18]. While active research continues into potential alternative neural net architectures like *state space models* [GD23; Fen+24] and *convolutional models* [Pol+23], the transformer architecture continues to dominate the field of text generation.

6.5.1 Attention mechanism

The word *transformer* nowadays is just a broad name for the overall multi-layer neural net structure used for text prediction. The core innovation in the architecture is the so-called **attention mechanism**. The idea behind the attention mechanism is that the function processing a sequence of token-related embeddings should be flexible enough and should depend on the content of the embeddings themselves – the simple average function that we used earlier just does not cut it as it destroys statistical information in the tokens. Attention, as a functional form, addresses this consideration.

```
# toy example
torch.manual_seed(999)
```

```
B = 4 # batch size (number of sequences)
S = 6 # sequence length
E = 2 # embedding vector length

# random batch - sequences of embeddings
X_emb = torch.randn(B, S, E)

print("Random batch:", X_emb)
print("Batch shape:", X_emb.shape)


Random batch: tensor([[[-0.2528,  1.4072],
         [ 0.2910,  1.0365],
         [-0.9816, -3.4219],
         [ 1.4910,  0.2422],
         [ 1.4832, -0.3704],
         [ 0.0941,  2.1528]],

        [[ 0.6271, -1.1666],
         [-0.7862,  0.0759],
         [-0.0086, -0.6568],
         [-1.0011,  0.2992],
         [ 0.6396, -1.0857],
         [-1.6153,  1.5635]],

        [[-1.7952,  0.6095],
         [-0.7203,  0.6119],
         [ 0.3259, -1.6059],
         [-0.5272,  0.3401],
         [-1.3832,  1.1149],
         [-0.7776,  0.2738]],

        [[ 0.9147, -1.1896],
         [-0.7501, -1.5465],
         [ 1.0044, -0.0986],
         [ 1.3962, -0.9138],
         [-1.1788, -0.6681],
```

```
          [-0.3168,  0.9893]]])
Batch shape: torch.Size([4, 6, 2])
```

As in the NumPy example earlier, each of 4 blocks above represents a separate sequence of 6 tokens, each token represented by an embedding of two scalars.

In our simple text prediction model, we just took an average of the embeddings within each sequence, getting an averaged embedding for each sequence in a batch, so $B \times S \times E$ arrays becomes $B \times E$:

```
X_emb_avg = torch.mean(X_emb, dim=1) # B x E
X_emb_avg
```

```
tensor([[ 0.3542,  0.1744],
        [-0.3575, -0.1618],
        [-0.8129,  0.2240],
        [ 0.1783, -0.5712]])
```

Note, that we can also write the exact same average computation in a few different ways (this will become useful later):

```
# weights - vector of 6 scalars of 1/6
w = torch.ones(S) / S

# dot product
X_emb_avg = w @ X_emb
X_emb_avg

# due to PyTorch broadcasting along batch dimension:
# S @ B x S x E >>> B x S @ B x S x E >>> B x E
```

```
tensor([[ 0.3542,  0.1744],
        [-0.3575, -0.1618],
        [-0.8129,  0.2240],
        [ 0.1783, -0.5712]])
```

As you can see, the result is equivalent – due to some PyTorch tensor broadcasting magic. And now another rewrite:

```
w = torch.ones(S)

# softmax - gives equal probabilities
w = torch.exp(w)
w = w / w.sum()

# broadcasting magic again
# S @ B x S x E  >>>  B x E
X_emb_avg = w @ X_emb
X_emb_avg
```

```
tensor([[ 0.3542,  0.1744],
        [-0.3575, -0.1618],
        [-0.8129,  0.2240],
        [ 0.1783, -0.5712]])
```

Because the weight vector w has all elements equal to each other, pushing it through softmax yields equal (uniform) probabilities – and thus a regular average.

Attention mechanism introduced a key idea – such weights should not be set arbitrarily – they should be set by the neural net itself, in a learnable way, as a function of inputs. The specific mechanism is as follows:

1. Define two $E \times m_1$ matrices W_q, W_k and the $E \times m_2$ matrix W_v – these are trainable parameters – you can think about them as three different linear layers of neurons without intercept parameters.

2. Based on input X_{emb}, which is shaped $B \times S \times E$, compute three matrices:

$$\text{Query:} \quad X_q = X_{emb} \cdot W_q;$$
$$\text{Key:} \quad X_k = X_{emb} \cdot W_k;$$
$$\text{Value:} \quad X_v = X_{emb} \cdot W_v.$$

3. Compute

$$\underset{B \times S \times S}{\Sigma} = \underset{B \times S \times m_1}{X_q} \cdot \underset{B \times m_1 \times S}{X_k^T} / \sqrt{m_1}$$

via broadcasting along the batch dimension, where Σ can be thought of as covariance, for each sequence in a batch, between token embeddings, scaled by square root of query / key dimension m_1 (to ensure probabilities are spread out and not too peaked).

4. Convert Σ to row-wise probabilities (weights W) via softmax on the last dimension:

$$\underset{B \times S \times S}{W} = \text{softmax}\left(\underset{B \times S \times S}{\Sigma}, \dim = -1\right).$$

Notice that now the weight tensor describes for each token how much "attention" it should pay to every other token in the sequence, which gives the name to the *attention mechanism*.

5. Compute

$$\underset{B \times S \times m_2}{X_{\text{out}}} = \underset{B \times S \times S}{W} \cdot \underset{B \times S \times m_2}{X_v},$$

where, via broadcasting, for each element in a batch, dot product happens between $S \times S$ and $S \times m$ arrays in W and X_v respectively.

Altogether, the *attention function* is then this:

$$
\begin{aligned}
X_{\text{out}} &= f_{\text{attn}}(X_{\text{emb}}) \\
&= \text{softmax}_{\dim=-1}(X_q \cdot X_k^T / \sqrt{m_1}) \cdot X_v \\
&= \text{softmax}_{\dim=-1}(X_{\text{emb}} \cdot W_q \cdot W_k^T \cdot X_{\text{emb}}^T / \sqrt{m_1}) \cdot X_{\text{emb}} \cdot W_v,
\end{aligned}
$$

with W_q, W_k, W_v as its trainable parameters.

What the attention function does is it creates a new sequence of embeddings based on a past sequence of embeddings, where each new embedding is a mixture (weighted average / linear transformation) of past embeddings. The mixture proportion is determined by the data and the model.

This transformation turns out to be incredibly powerful and expressive. This is akin to a simple average of embeddings we took in the prior simple model – but instead of one average, we compute multiple averages, the averages are weighted in a data-dependent way, and we are averaging not raw inputs, but inputs transformed via linear transformation (dot product with a matrix).

Here is a basic **attention** implementation:

```
torch.manual_seed(999)

B, S, E = 4, 6, 2

# input data
X_emb = torch.randn(B, S, E)

# attention parameters
m_1 = m_2 = E # setting key/query and value matrix sizes
W_q = torch.randn(E, m_1)
W_k = torch.randn(E, m_1)
W_v = torch.randn(E, m_2)

# attention forward computation
X_q = X_emb @ W_q   # B x S x m_1
X_k = X_emb @ W_k   # B x S x m_1
X_v = X_emb @ W_v   # B x S x m_2

W = X_q @ X_k.transpose(-2, -1) # B x S x S
W = W / (m_1**0.5)
W = torch.softmax(W, dim=-1)

X_out = W @ X_v # B x S x m_2
print(X_out.shape)
```

```
torch.Size([4, 6, 2])
```

Notice also that if we set the value matrix dimension size $m_2 = E$, then shapes of X_{emb} and X_{out} will be the same at $B \times S \times E$. So we could keep passing this tensor on and on and on through attention, getting more and more complex transformations, while the dimensionality of the array stays the same. This array can then be converted to next-token probabilities for each sequence at the output via a simple linear layer and softmax.

Alternatively, we could, for example, make $m_2 = E/2$ – as long as m_2 stays integer – and then run 2 attention computations in parallel – and then concatenate their outputs along the embedding dimension – this would also result in the same output shape E – but now with different parameters operating on input and generating different parts of the output – this is called a *multi-head attention*.

Also notice that sequences in a batch are totally independent here – their data in no way gets mixed through this forward computation. In fact, we really only include the batch dimension here because it helps run the computation in parallel efficiently.

This attention function – with some modifications we will see next – is the foundation of the transformer architecture, which applies the attention transformation mixed with some linear layers again and again on large scale.

The presented attention function is also called *self-attention* because arrays X_v, X_q, X_k are all ultimately derived from the same input sequence. This allows the model to learn relationships within that sequence – how one token relates to others in a sentence. In contrast, *cross-attention* uses different input sequences to compute X_q vs. X_v and X_k. This enables a model to relate information across sequences – such as aligning a translation in one language to its source in another. In this chapter, we focus exclusively on self-attention. So unless otherwise noted, "attention" always refers to self-attention.

6.5.2 More efficient data utilization

Before we jump into implementing the full transformer model in PyTorch – where its details will be shown in code – we need to tackle a slight inefficiency in how we handle input and output data during training.

In the computation so far, we have been predicting the next (hold-out) character – and all inputs used in the computation above precede it – so we can mix these inputs together via this attention function without worrying about somehow leaking information into the future.

Recall that our existing function to get the batch gives us a $B \times S$ array X_{ind} and a $B \times 1$ array y_{ind} – each separate row is an input sequence of character indices of length S contained in X_{ind} and the $S+1$ character index contained in y_{ind}.

```
torch.manual_seed(999)
get_batch('train')
```

```
(tensor([[46, 44, 57, 75, 63,  2],
         [61, 62, 48, 55, 49,  1],
         [62,  2, 44, 62,  2, 62],
         [77,  2, 16, 55, 52, 46]], device='mps:0'),
 tensor([63,  0, 63, 48], device='mps:0'))
```

Training with this batch requires us to have S characters ready before we can start predicting. However, we want to already start predicting from 1, 2, 3, and so on characters – furthermore, input array X_{ind} may already have targets for those predictions – for example, if we are using the first two characters for input, trying to predict the third character, and X_{ind} has sequences of length 6, then third column in X_{ind} could play the role of y_{ind} for the input of the first two columns. So, if properly shifted, X_{ind} can serve as its own prediction target – but instead of just one type of prediction target it would contain a multitude of targets – for input sequences of increasing length.

Let us write it out in code as a new updated **get_batch** function, with a new definition of the output Y_{ind}:

```
torch.manual_seed(999)

B = 4    # number of token sequences
S = 6    # input sequence length

def get_batch(split):
    # get text sequences from random starting points
    if split == 'train':
        d = data_train
    if split == 'test':
        d = data_test
```

```
    start_ind = torch.randint(d.shape[0] - S, (B,))
    X_ind = torch.stack([d[i:i+S] for i in start_ind])

    # NOTE THE CHANGE:
    Y_ind = torch.stack([d[i+1:i+S+1] for i in start_ind])

    X_ind, Y_ind = X_ind.to(device), Y_ind.to(device)
    return X_ind, Y_ind

X_ind, Y_ind = get_batch('train')

print("X_ind:",X_ind)
print("Y_ind:",Y_ind)


X_ind: tensor([[46, 44, 57, 75, 63,  2],
        [61, 62, 48, 55, 49,  1],
        [62,  2, 44, 62,  2, 62],
        [77,  2, 16, 55, 52, 46]], device='mps:0')
Y_ind: tensor([[44, 57, 75, 63,  2, 63],
        [62, 48, 55, 49,  1,  0],
        [ 2, 44, 62,  2, 62, 63],
        [ 2, 16, 55, 52, 46, 48]], device='mps:0')
```

As can be seen, if an input sequence starts at some index i in text, and is of length S, then the output sequence will also be of length S, but starting at index $i + 1$. Then the first element in the output sequence is a valid next-character prediction target for the first element in input sequence. The second element in the output sequence is a valid next-character prediction target for the first two elements in input sequence. And so on.

```
X_ind[0,:3] # first 3 elements in the first sampled sequence
Y_ind[0,3]  # 4th element in that sequence, a valid
 ↪  prediction target for the first sequence
X_ind[0,4]  # this is the same as Y_ind[0,3]


tensor(63, device='mps:0')
```

Please make sure you understand this somewhat confusing structuring of data. Its purpose is to make data utilization more efficient.

6.5.3 Causal attention

The last ingredient that is needed here is a filter to ensure that when attention function is applied and predicts, for instance, the third character, it can use only information from the embeddings of the first two characters – and not the embeddings of characters 3 and later – because that would be using information from the future of the string to predict the future string – and we do not want to allow this data leakage. (Such peeking into the future may be fine for a different type of task, for example, translation, but we won't discuss it here.)

We will implement such *causal* filter using some matrix algebra trickery.

Example batch:

```
torch.manual_seed(999)
B, S, E = 4, 6, 2  # batch size, sequence length, embd size
X_emb = torch.randn(B, S, E)
X_emb.shape
```

```
torch.Size([4, 6, 2])
```

```
X_emb[0] # first sequence of embeddings
```

```
tensor([[-0.2528,  1.4072],
        [ 0.2910,  1.0365],
        [-0.9816, -3.4219],
        [ 1.4910,  0.2422],
        [ 1.4832, -0.3704],
        [ 0.0941,  2.1528]])
```

Consider two ways to take an average of embeddings up to a point in a sequence.

(a) Using a for loop:

```
for i in range(S):
    print(torch.mean(X_emb[0][:i+1,:],0))
```

```
tensor([-0.2528,  1.4072])
tensor([0.0191, 1.2218])
tensor([-0.3144, -0.3261])
tensor([ 0.1369, -0.1840])
tensor([ 0.4062, -0.2213])
tensor([0.3542, 0.1744])
```

(b) Using a lower triangular matrix:

```
# initialize the lower triangular matrix
tril = torch.tril( torch.ones(S, S) )
tril
```

```
tensor([[1., 0., 0., 0., 0., 0.],
        [1., 1., 0., 0., 0., 0.],
        [1., 1., 1., 0., 0., 0.],
        [1., 1., 1., 1., 0., 0.],
        [1., 1., 1., 1., 1., 0.],
        [1., 1., 1., 1., 1., 1.]])
```

```
# normalize its rows to sum to 1
W = tril / tril.sum(1, keepdim=True)
W
```

```
tensor([[1.0000, 0.0000, 0.0000, 0.0000, 0.0000, 0.0000],
        [0.5000, 0.5000, 0.0000, 0.0000, 0.0000, 0.0000],
        [0.3333, 0.3333, 0.3333, 0.0000, 0.0000, 0.0000],
        [0.2500, 0.2500, 0.2500, 0.2500, 0.0000, 0.0000],
        [0.2000, 0.2000, 0.2000, 0.2000, 0.2000, 0.0000],
        [0.1667, 0.1667, 0.1667, 0.1667, 0.1667, 0.1667]])
```

```
# dot product to get the averages
W @ X_emb[0]
```

```
tensor([[-0.2528,  1.4072],
        [ 0.0191,  1.2218],
        [-0.3144, -0.3261],
        [ 0.1369, -0.1840],
        [ 0.4062, -0.2213],
        [ 0.3542,  0.1744]])
```

Compare this result to the for loop result – they are identical.

Now, one more trick. Say we have an arbitrary matrix of weights W. We can set parts of it to minus infinity based on **tril** matrix – for those elements in sequences we do not want to use in the computation – to avoid data leakage. Such negative infinity weights then become 0 after softmax. If elements of the weight matrix are initially equal, we get a simple average, as before:

```
W = torch.ones(S, S) # arbitrary weight matrix
mask = torch.tril(torch.ones(S, S))
W = W.masked_fill(mask == 0, float('-inf'))
W
```

```
tensor([[1., -inf, -inf, -inf, -inf, -inf],
        [1., 1., -inf, -inf, -inf, -inf],
        [1., 1., 1., -inf, -inf, -inf],
        [1., 1., 1., 1., -inf, -inf],
        [1., 1., 1., 1., 1., -inf],
        [1., 1., 1., 1., 1., 1.]])
```

```
W = torch.softmax(W, dim=-1)
W
```

```
tensor([[1.0000, 0.0000, 0.0000, 0.0000, 0.0000, 0.0000],
```

```
             [0.5000, 0.5000, 0.0000, 0.0000, 0.0000, 0.0000],
             [0.3333, 0.3333, 0.3333, 0.0000, 0.0000, 0.0000],
             [0.2500, 0.2500, 0.2500, 0.2500, 0.0000, 0.0000],
             [0.2000, 0.2000, 0.2000, 0.2000, 0.2000, 0.0000],
             [0.1667, 0.1667, 0.1667, 0.1667, 0.1667, 0.1667]])
```

```
W @ X_emb[0]
```

```
tensor([[-0.2528,  1.4072],
        [ 0.0191,  1.2218],
        [-0.3144, -0.3261],
        [ 0.1369, -0.1840],
        [ 0.4062, -0.2213],
        [ 0.3542,  0.1744]])
```

The initial weight matrix W can be arbitrary, e.g., determined by the model, allowing us to take weighted averages of the embeddings in a data-dependent way.

Now we need to build this idea into the attention function – so every next element in a sequence can only attend to / be computed based on elements that precede it in time. This modified function is called **causal attention** and here is its implementation:

```
torch.manual_seed(999)

B, S, E = 4, 6, 2

# input data
X_emb = torch.randn(B, S, E)

# attention parameters
m_1 = m_2 = E # setting key/query and value matrix sizes
W_q = torch.randn(E, m_1)
W_k = torch.randn(E, m_1)
W_v = torch.randn(E, m_2)
```

```
# attention forward computation
X_q = X_emb @ W_q   # B x S x m_1
X_k = X_emb @ W_k   # B x S x m_1
X_v = X_emb @ W_v   # B x S x m_2

W = X_q @ X_k.transpose(-2, -1)  # B x S x S
W = W / (m_1**0.5)

# causal filter

# (S, S) lower triangular mask
mask = torch.tril(torch.ones(S, S))
W = W.masked_fill(mask == 0, float('-inf'))
W = torch.softmax(W, dim=-1)

# the mask needs to be broadcasted to (B, S, S) if necessary
# where mask is 0, set to negative infty >>> 0 after exp()

X_out = W @ X_v  # B x S x m_2
print(X_out.shape)
```

```
torch.Size([4, 6, 2])
```

Notice that now each output embedding-like vector of length m_2 is computed strictly based on inputs at its position in a sequence or before it – and no information from ahead of it gets used.

6.5.4 Transformer architecture based on causal self-attention

We will now implement a new neural net structure and training based on this attention mechanism for a relatively small LLM. We will be using causal self-attention, but will simply refer to it as attention.

In addition to attention, you will see some special tricks in the code that have been found to make the training process for the transformer function go smoothly and efficiently. These include:

- *LayerNorm* [BKH16] is an operation that normalizes an array across its last dimension to zero mean and unit variance – in our case, across the components of each embedding vector – using the computed mean and standard deviation. After that, the normalized output is scaled and shifted using *learned* scale and bias parameters. In effect, this replaces empirical means and standard deviations with learned ones. This procedure has been found to help stabilize training.

- *Skip connections* (also known as residual connections) allow the input of a layer to bypass one or more intermediate layers and be added directly to their output. This design helps preserve signal during backpropagation, making it easier to train deep networks by mitigating the *vanishing gradient* problem – a situation where gradients used to update weights during backpropagation can become very small as they are propagated backward through very deep networks [He+16; Li+18].

- *Position embeddings* are special learned embeddings – one per possible position in the token sequence [Dev+19; Vas+17]; you can think of these as a set of position-specific vector-shaped bias / intercept terms. These can help the model distinguish the order of tokens.

The model that is presented here largely follows a reduced-in-size model architecture from OpenAI's GPT-2 paper [Rad+19]. In turn, GPT 2 architecture almost exactly follows GPT-1 architecture [Rad+18] – with such minor changes as LayerNorm placement. In turn, GPT-1 architecture mimics relatively closely the *decoder* (causal) module in the original attention paper [Vas+17], but uses self-attention instead of cross-attention due to the shift from language translation settings to the straightforward new text generation.

Note that we do **not** use *weight tying* [PW16] – the sharing of the vocabulary array of the token embedding weights both (a) for encoding input text as well as (b) in the role of weights in the final output layer that computes logits. Weight tying was used in GPT-2 paper. This technique reduces the number of parameters and may improve training efficiency, but at the cost of the model's flexibility. Given that our vocabulary is relatively small, the model could perhaps benefit from more flexibility, so we will learn embedding weights and output linear layer weights separately.

The *transformer*, in words, is a neural net consisting of a list of architecturally identical *transformer blocks*, applied sequentially, with the final output linear layer at the end of transformer mapping the output of the list of transformer blocks to the predicted logits over possible next tokens. These logits can be used to compute token probabilities.

Each transformer block consists of a *multi-head attention* module followed by a *feed forward* module. Multi-head attention is just several (n_{heads}) *attention modules / heads* running in parallel, their outputs concatenated along the outputs' last dimension (i.e., embedding dimension) – plus a linear layer on top. Each attention head outputs a mini-embedding of size E/n_{heads}, so – after concatenation – the last dimension of the multi-head attention output array is E. Feed forward module is just a couple of linear layers one after another.

Each transformer block involves layer normalization of inputs prior to each *multi-head attention* module and *feed forward* module steps. Additionally, each transformer block involves skip connections.

There is also a layer normalization before the final output linear layer in the transformer. Throughout the transformer model, we occasionally use dropout (randomly setting neuron outputs to zero, for regularization) and ReLU non-linearity (GPT-2 uses a more sophisticated GELU activation [HG16], but ReLU is good enough for our purposes).

It is much better and less ambiguous though to see the transformer's implementation in code! You will see that the code uses different custom classes for a clean decomposition. I will provide a brief explanation where necessary, though the code should be mostly self-explanatory.

We first define some parameters:

```
B = 64 # batch size -- number of sequences
S = 256 # sequence length, tokens per sequence
E = 64 # embedding size per token

n_heads = 4 # number of parallel attention heads
n_layers = 4 # number of transformer blocks
dropout = 0.2 # dropout probability (regularization)
```

Now we create the basic causal self-attention class:

```python
class Attention(nn.Module):
    # causal self-attention module

    def __init__(self, m):
        super().__init__()
        self.m = m # m_1 = m_2

        self.key = nn.Linear(E, m, bias=False) # W_k
        self.query = nn.Linear(E, m, bias=False) # W_q
        self.value = nn.Linear(E, m, bias=False) # W_v

        # create a non-trainable lower triangular matrix
        self.register_buffer('tril',
            torch.tril(torch.ones(S, S)))
        self.dropout = nn.Dropout(dropout) # regularization

    def forward(self, X_emb):
        B, S, m = X_emb.shape

        X_k = self.key(X_emb)    # B x S x m
        X_q = self.query(X_emb) # B x S x m
        X_v = self.value(X_emb) # B x S x m

        # B x S x m @  B x m x S  >>>  B x S x S
        W = X_q @ X_k.transpose(-2,-1) / self.m**0.5

        W = W.masked_fill(self.tril[:S, :S] == 0,
                float('-inf'))    # B x S x S
        W = F.softmax(W, dim=-1) # B x S x S
        W = self.dropout(W)

        # B x S x S @  B x S x m  >>>  B x S x m
        X_out = W @ X_v
        return X_out
```

We now define a multi-head attention class, containing a layer of parallel attention functions, finished up with a linear layer and dropout regularization:

```
class MultiHeadAttention(nn.Module):
    # group of attention heads running in parallel

    def __init__(self, n_heads, head_size):
        super().__init__()
        self.heads = nn.ModuleList([
            Attention(head_size) for i in range(n_heads)
        ])
        self.lin = nn.Linear(E, E)
        self.dropout = nn.Dropout(dropout)

    def forward(self, X_emb):
        X_out = torch.cat([
            h(X_emb) for h in self.heads
        ], dim=-1) # B x S x E
        X_out = self.dropout(self.lin(X_out))
        return X_out
```

We now define the feed forward module, which follows the multi-head attention module in a transformer block. The feed forward module is just a neural net with two fully connected layers of the following form:

```
class FeedForward(nn.Module):
    # neural net of two fully connected layers

    def __init__(self, E):
        super().__init__()
        self.layers = nn.Sequential(
            nn.Linear(E, 4*E),
            nn.ReLU(),
            nn.Linear(4*E, E),
            nn.Dropout(dropout))

    def forward(self, X_emb):
        return self.layers(X_emb)
```

Here is the LayerNorm implementation for layer normalization:

```python
class LayerNorm(nn.Module):
    # layer normalization

    def __init__(self, D, eps=1e-5):
        super().__init__()
        self.eps = eps
        self.scale = nn.Parameter(torch.ones(D))
        self.bias = nn.Parameter(torch.zeros(D))

    def forward(self, x):
        mean = x.mean(dim=-1, keepdim=True)
        var = x.var(dim=-1, unbiased=False, keepdim=True)
        x_norm = (x - mean) / torch.sqrt(var + self.eps)
        return self.scale * x_norm + self.bias
```

Putting these components together, we get the transformer block class:

```python
class TransformerBlock(nn.Module):
    # key transformer's building block

    def __init__(self, E, n_heads):
        super().__init__()
        # E and n_heads should be selected
        # so that E / n_heads has 0 remainder
        head_size = E // n_heads
        self.mha = MultiHeadAttention(n_heads, head_size)
        self.ffwd = FeedForward(E)
        self.lnorm1 = LayerNorm(E)
        self.lnorm2 = LayerNorm(E)

    def forward(self, X_emb):
        # skip connections via addition of input to output
        X_emb = X_emb + self.mha(self.lnorm1(X_emb))
        X_emb = X_emb + self.ffwd(self.lnorm2(X_emb))
        return X_emb
```

Finally, here is our crown jewel, the transformer model based on causal self-attention, consisting of a sequence of transformer blocks, topped with layer normalization and a fully connected linear output layer.

This is our basic LLM.

```python
class Transformer(nn.Module):
    # a basic transformer architecture LLM

    def __init__(self, V, E):
        super().__init__()

        # an embedding for each token
        self.token_embeddings = nn.Embedding(V, E)

        # and an embedding for each sequence position
        self.position_embeddings = nn.Embedding(S, E)

        # sequence of repeated transformer blocks
        self.blocks = nn.Sequential(*[TransformerBlock(E,
            n_heads=n_heads) for i in range(n_layers)
        ])

        # output layer
        self.output_lnorm = LayerNorm(E)
        self.output_layer = nn.Linear(E, V)

    def forward(self, X_ind):
        B, S = X_ind.shape
        tok_emb = self.token_embeddings(X_ind) # B x S x E
        pos_emb = self.position_embeddings(
            torch.arange(S, device=device)) # S x E
        X_emb = tok_emb + pos_emb # B x S x E
        X_emb = self.blocks(X_emb) # B x S x E
        X_emb = self.output_lnorm(X_emb) # B x S x E
        X_logits = self.output_layer(X_emb) # B x S x V
        return X_logits

    def loss(self, X_logits, Y_ind):
```

```
    # X_logits is B x S x V; Y_ind is B x S
    return F.cross_entropy(
                X_logits.view(-1, X_logits.shape[-1]),
                Y_ind.view(-1))

def next_token_prob(self, X_logits):
    # X_logits is B x S x V
    X_logits = X_logits[:, -1, :] # B x V
    X_prob = F.softmax(X_logits, dim=-1) # B x V
    return X_prob
```

Finally, we will use a helper function to initialize the transformer's weights to appropriate values, which helps speed up training:

```
def init_weights(module):
    if isinstance(module, (nn.Linear, nn.Embedding)):
        nn.init.normal_(module.weight, mean=0.0, std=0.02)
        if hasattr(module, 'bias') and module.bias is not
        ↪ None:
            nn.init.zeros_(module.bias)
    elif isinstance(module, LayerNorm):
        nn.init.ones_(module.scale)
        nn.init.zeros_(module.bias)
```

There are also more sophisticated approaches available for weight initialization that may deliver more efficient training – see, for example, sparse transformers [Chi+19].

Initialize the model:

```
torch.manual_seed(999)
net = Transformer(V, E)
net.apply(init_weights)
net = net.to(device)
```

Number of model parameters:

```
print(sum(p.numel() for p in net.parameters()))/1e6, 'million
 ↪  parameters')
```

```
0.225742 million parameters
```

Training loop:

```
max_iters = 30000
eval_interval = 3000
learning_rate = 1e-4 # 0.0001

optimizer = torch.optim.AdamW(net.parameters(),
    lr=learning_rate, weight_decay=0.1)

start_time = time.time()
for i in range(max_iters):

    # evaluation
    if i % eval_interval == 0 or i == max_iters - 1:
        # elapsed time in seconds
        elapsed = time.time() - start_time
        # reusing the eval function from earlier
        eval_ = evaluate(net)
        print(f"Step {i}: "
            f"Train loss = {eval_['train']:.4f}, "
            f"Test loss = {eval_['test']:.4f}, "
            f"\n\t Elapsed time = {elapsed/60:.2f} mins, "
            f"{100*(i+1)/max_iters:.0f} % complete\n")

    # training
    X_batch, Y_batch = get_batch('train')
    X_logits = net(X_batch)
    loss = net.loss(X_logits, Y_batch)
    optimizer.zero_grad()
    loss.backward()
    optimizer.step()
```

```
Step 0: Train loss = 4.3803, Test loss = 4.3808,
    Elapsed time = 0.00 mins, 0 % complete

Step 3000: Train loss = 1.6440, Test loss = 1.7307,
    Elapsed time = 5.19 mins, 10 % complete

Step 6000: Train loss = 1.2971, Test loss = 1.5063,
    Elapsed time = 9.55 mins, 20 % complete

Step 9000: Train loss = 1.1784, Test loss = 1.4534,
    Elapsed time = 13.89 mins, 30 % complete

Step 12000: Train loss = 1.1093, Test loss = 1.4398,
    Elapsed time = 18.21 mins, 40 % complete

Step 15000: Train loss = 1.0628, Test loss = 1.4304,
    Elapsed time = 22.49 mins, 50 % complete

Step 18000: Train loss = 1.0273, Test loss = 1.4260,
    Elapsed time = 26.84 mins, 60 % complete

Step 21000: Train loss = 1.0010, Test loss = 1.4223,
    Elapsed time = 31.12 mins, 70 % complete

Step 24000: Train loss = 0.9772, Test loss = 1.4253,
    Elapsed time = 35.43 mins, 80 % complete

Step 27000: Train loss = 0.9587, Test loss = 1.4209,
    Elapsed time = 39.76 mins, 90 % complete

Step 29999: Train loss = 0.9412, Test loss = 1.4215,
    Elapsed time = 44.12 mins, 100 % complete


print("Final test set cross-entropy (CE):",
    float(f"{eval_['test']:.4f}"))
print("Probability of picking the correct next token: ")
print("> Neural net (exp(-CE)):",
```

```
    float(f"{torch.exp(-eval_['test']):.4f}"))
print("> Uniform random guess (1/V):", round(1/V,4))
```

```
Final test set cross-entropy (CE): 1.4215
Probability of picking the correct next token:
> Neural net (exp(-CE)): 0.2413
> Uniform random guess (1/V): 0.0128
```

Generate text:

```
net.eval()  # turns off dropout
s = generate_text(net, n_new_tokens=500)
print(textwrap.fill(s, 60))
```

"Exactly," For said Alice; "but what to know."   And what
people half is Canto much happening and from like, but it
was it back out out like to fall on, "I' ll at they but her
way I do is to my hear now."   "See is their right the right
to she left!" exclas she King the Queen, turning at here
way good outt.   "The exest of a shome gimmpent?" said
Alice, aloud: "what o the Queen into of reaces."   "Yes, as
you might to livery good round for to that do," the Lory its
make out of she had not sai

Boom! As we can see, after just under 45 minutes of training on my Apple MacBook Pro M1 Max, we are getting much better results – including recognizable text.

We can now save the raw weights of the LLM as follows:

```
# only saving weights / parameter values
PATH = "./llm_weights.pth"
torch.save(net.state_dict(), PATH)
```

We can load the model weights back into a pre-defined LLM object as follows:

```
net = Transformer(V, E)
net.load_state_dict(torch.load(PATH,
    map_location=device, weights_only=True))
net = net.to(device)
net.eval()
```

Therefore, the LLM is effectively fully determined by 2 text files: the first text file contains the learned parameters; the second text file contains the class definition of the LLM architecture / the forward computation in Python code – the corresponding object can load in the weights from the first parameter file.

> ⚠ Why save just the weights and not the entire LLM?
>
> Saving the entire LLM via `torch.save(net, PATH)` is generally *discouraged*. This mechanism relies on Python's `pickle` library and it can break, for example, due to PyTorch version differences or even slight class refactoring, resulting in the inability to reload the LLM. Storing just the raw weights via `torch.save(net.state_dict(), PATH)` is more flexible and robust, as it decouples the model's learned weights from the class implementation.

6.6 Towards a commercial-grade LLM

The basic code we have for the causal-attention transformer underlies most modern industry-grade LLMs. However, quite a bit more work would need to be done to bring our model up to a level where it matches the top-performing models. In this section, we will fill in more details on how industry-grade models are actually crafted.

We need to acknowledge that, for many top LLMs, especially their more recent versions, training and architecture details might not be publicly available; only the older LLM iterations in the series might come with relatively detailed descriptions. OpenAI's models are a good example of this, where there are more details available on GPT-1 through GPT-3 architectures, but public information on GPT-4o, o1, o3, o4-mini models, etc. is far scarcer.

Luckily, there are some powerful LLMs, where their creators chose the more open-source approach and published not only detailed descriptions of the training procedure, but also released the weights of the models to the public – so the architecture is transparent. Examples of popular and powerful open-weight LLMs, by jurisdiction of the creator company, include:

- China:
    - DeepSeek-V3 [Liu+24] and DeepSeek-R1 [Guo+25].
    - Kimi K2 [Kim25] by Moonshot AI.
    - MiniMax-M1 [Che+25] by MiniMax AI.
    - Qwen 2.5 [Yan+25a] and Qwen 3 [Yan+25b] by Alibaba Cloud.

- France:
    - Mistral [Jia+23].

- USA:
    - Llama 3 [Gra+24] and Llama 4 [Met25] by Meta.

To get a ranking of these models in terms of performance on different benchmarks based on paired-comparison human feedback, you can visit LMArena website (https://lmarena.ai/leaderboard) [Chi+25]. SEAL LLM Leaderboards rank models by performance on a set of benchmarks (https://scale.com/leaderboard). The technical reports that often come out hand-in-hand with the LLM release (linked in the list above) also contain comparisons on different benchmarks across a set of selected competing models.

To give a brief historical overview, open-weight models like Qwen 2.5 and Llama 3 came out in 2024. Then, towards the end of 2024, DeepSeek-V3 model came out and proved to be a top-performing state-of-the-art open-weight model at the time, beating Qwen 2.5 and Llama 3 on many benchmarks – it was even comparable to closed-weight Anthropic's Claude 3.5 Sonnet [Ant24b] and OpenAI's GPT-4o [Hur+24]. (As we will discuss later, DeepSeek-V3 also turned out to be embarrassingly cheap to train compared to the competitors.) DeepSeek-R1 is a model based on DeepSeek-V3 and optimized for multi-step reasoning.

Later, in April 2025, Qwen 3 was released, which built on DeepSeek's methodology and improved on base DeepSeek models in performance; Llama 4 came out around the same time, but showed somewhat disappointing performance, underperforming Qwen 3 [Yan+25b] – this is likely what triggered a subsequent attempt within Meta to build a new AI team [Blo25].

Then, in July 2025, Kimi K2 was released by Moonshot AI, apparently beating, on average, all open-weight models in performance and being close to top commercial models like Anthropic's Claude Opus 4 [Kim25; Ant25]. Kimi K2 also heavily built on DeepSeek's architectural decisions and can be thought of as its scaled up version.

As of August 2025, commercial closed-weight models like OpenAI's latest iteration of GPT-4o model (and newer GPT-5), Google's Gemini 2.5 Pro, and X's Grok 4 lead in the LLM LMArena rankings, but open-weight Kimi K2 and some updated versions of DeepSeek are close behind. It is worth giving credit where it is due – China is currently really doing a great job with the open-source LLM research and is leading the way vs. its foreign counterparts. Additionally, as an apparent response to the open-source developments, OpenAI released its open-weight `gpt-oss` series in August 2025 [Ope25a]; however, it did not rank very high on LMArena leaderboard as of the post-release evaluation.

We will now review some key areas, where more work is needed, compared to our basic code, to match the industry-grade LLMs in quality and scale.

6.6.1 More pre-training data

Industry-grade LLMs are trained in stages, which, at the high level, include (1) *pre-training* and (2) *fine-tuning*.

At scale, pre-training means optimizing the LLM's parameters to predict the next token from prior context on enormous, noisy, multi-domain text corpus – covering scraped web pages, PDFs, research papers, books, code, etc. – effectively, training the LLM to be an auto-completion engine. Pre-training phase is akin to what we have implemented in code so far on the tiny *Alice in Wonderland* text.

For illustration, Llama 1 paper [Tou+23b] discloses the composition of their pre-training data (close to 5 terabytes):

- CommonCrawl repository of scraped internet data: https://common crawl.org/.
- C4 [Raf+19] – CommonCrawl's cleaned version: https://www.tensor flow.org/datasets/catalog/c4.
- Wikipedia: https://www.wikipedia.org/.
- Books.
- arXiv preprint website: https://arxiv.org/.
- Stack Exchange: https://stackexchange.com/.

In Llama 1, the combined pre-training text was around 1.4 trillion (1.4T) tokens long; with $V = 32000$ (32K) unique tokens in the vocabulary [Hum23]. Tokenization was done via byte pair encoding (BPE), which we discussed in detail earlier in the chapter.

> **i** Average token
>
> A token used by the current leading LLMs is commonly taken to be, on average, around 4 English characters long – or around 0.75 words [Gra+24; Ope25c].

For other open-weight models, the (approximate) token numbers for (a) the length of training corpus, and (b) the vocabulary size are as follows:

- DeepSeek-V3: 14.8T / 128K [Liu+24].
- Kimi K2: 15.5T / 160K [Kim25].
- Llama 3: 15.6T / 128K [Gra+24].
- Llama 4: 30T / 202048 [Met25].
- Qwen 2.5: 18T / 151643 [Yan+25a].
- Qwen 3: 36T / 151669. [Yan+25b].

All of these works used variations of BPE for tokenization. And all of them provide little detail on the exact pre-training text composition.

It is well established that the composition and quality of the pre-training corpus strongly influence the final LLM performance. Large effort is devoted, for instance, to (a) re-weighting sources (e.g., emphasizing higher-value academic prose over lower-signal social media or e-commerce text), (b) filtering or excluding low-quality passages via learned quality prediction models, and (c) enriching the data mix with additional code, mathematical, and other content – including synthetic data generated by existing LLMs

[Yan+25a; Gra+24; Fuj+25; Kim25]. The process of creating new data from existing data to train LLMs is called *data augmentation* [Wan+24a]. In large scale training, particular care might also be taken to prevent self-attention from crossing document boundaries within a sequence [Gra+24].

6.6.2 Sidebar: Pre-training data and copyright

Even for the open-weight models, the exact training data sets are often not released. To some degree, this helps AI teams maintain some competitive advantage. However, perhaps to a greater degree, this is done to mitigate possible legal liability in cases of the copyright-related legal action.

Currently, the legality of training LLMs on the publicly available copyrighted texts (including the pirated content) is a developing area of law in the United States, with active ongoing litigation.

The US courts in their recent decisions [Cala; Calb] seem to be leaning towards an LLM-friendly stance – that training LLMs on legally procured copyright-protected text is transformative enough to qualify LLM-generated content for fair-use treatment – even if the LLM training involves making incidental digital copies of the purchased physical books (and despite the lack of explicit permission from copyright holders for such copying).

The caveat is that such LLM content should not cause demonstrable market harm to the original copyright owner, e.g., by reproducing substantial portions of the source books. Luckily for LLM makers, it can be very hard to prove such specific monetary harm in practice. For instance, seeing a brief book passage reproduced exactly by an LLM might not hurt the book sales but instead boost them – consumers may decide to purchase the item after sampling it – so the mere fact of exact reproduction of the text does not automatically mean market harm.

There is, however, an apparent agreement that using pirated book copies for training is not protected by fair use – so the exact source of the training data remains an important consideration.

In practice though, up-and-coming AI labs tend not to worry too much about these legal issues, deciding they can ask for forgiveness later and using in LLM training whatever data they can get their hands on. The rationale for this is straightforward.

First, if an AI startup succeeds at creating a super strong LLM based on uncleanly sourced data, they likely will get copious amounts of funding, which then pays for any lawyers they might need; and if the startup fails to create a successful LLM, almost no one is going to use the said LLM (or even hear about it), so there are no real damages to copyright holders and only a minimal chance of copyright-related legal consequences.

Second, if a specific country decides to get strict with copyright enforcement and chooses to impede LLM development on these grounds – they risk pushing LLM developers to flee their jurisdiction for countries that do not care about copyright as much. Given that losing the international AI race might have bad national security consequences, nourishing domestic LLM models and AI talent thus seems to be a national security priority. And I would venture to think that national security concerns would trump copyright legalese in countries like the US or China.

Third, thinking from first principles, the primary purpose of the copyright law is to promote society's creative output. For example, in the US, this is written into the Constitution:

> "Congress shall have the Power ... To promote the Progress of Science and useful Arts, by securing for limited Times to Authors and Inventors the exclusive Right to their respective Writings and Discoveries."

> — United States Constitution, Article I, Section 8.

As in this example, current copyright frameworks tend to promote the creative output by incentivizing human creators with money, granting them exclusive rights to their intellectual property. However, we are now entering the world where AI can out-produce human creators, e.g., generating more quality images or code per unit of time than a human ever could – and AI can do it 24/7 without a need for rest or compensation.

If copyright laws start restricting AI creativity, such laws could then have a net negative effect on the overall society's creative output (especially, given that AI's share of this output is bound to keep growing). This would undermine the very reason for copyright's existence and could convince lawmakers to alter the copyright law to realign it with its original purpose of maximizing creative output – even if it means reduced income for human creators. AI labs might be counting on this.

6.6.3 Larger model size

Commercial-grade models are much larger in the number of parameters than our basic model – they include more transformer blocks, attention heads, embedding dimensions, etc. Their training requires the use of thousands of GPUs in parallel.

Compared to our model of < 1 million (1M) parameters, the largest GPT-3 model, for instance, has 175 billion (175B) parameters, 50257 tokens constructed via BPE, 12288 parameters in each token embedding vector, 96 attention heads in parallel within each multi-head attention module, and the sequence of 96 transformer blocks producing the computation [Bro+20].

Note that 175B parameters translate to roughly *350GB* of memory, assuming rounding of parameters to 16-bit floats – a single 16-bit float parameter occupies 2 bytes, so 175 billion 16-bit float parameters occupy $2 \cdot 175 = 350$ billion bytes, which is 350GB (1GB equals 1 billion bytes).

A top-line NVIDIA GPU like A100 has working memory of only *80GB*, so at least 5 GPUs would be needed to load in such whole model at the same time! Situation is further aggravated once you realize that for training these models, optimizers have to store extra variables, e.g., for gradient accumulation and gradient history tracking, which means even more GPU demand.

A GPU like A100 currently costs tens of thousands of US dollars. GPUs also have a tendency to break relatively fast – some estimates point to as low as 1-3 years of expected lifespan under heavy utilization [Shi24]. All this explains the skyrocketing stock prices of GPU companies [Lie25].

To give you a sense of the scale of GPU demand, training of Llama 1 largest 65B-parameter model involved processing 380 tokens/sec/GPU on *2048* A100 GPUs with 80GB of RAM; the training over 1.4T-token data set took approximately 21 days [Tou+23b].

And some of the more recent LLMs are even larger. For instance, Llama 3 largest model had 405B parameters, 128K tokens in the vocabulary, embedding size 16384, 128 attention heads, and 126 transformer blocks; it was trained on 15.6T tokens. Llama 3 405B was trained on up to *16K* H100 GPUs, the pre-training taking 54 days [Gra+24]. This means much more GPU utilization compared to Llama 1.

Interestingly, the even more recent DeepSeek-V3, due to some optimizations that we mention later, such as mixture-of-experts architecture and low-precision training, only required training for around two months on *2048* H100 GPUs at the strikingly low reported total rental GPU-compute cost of USD 5.576M for the training run [Liu+24]. Yet, DeepSeek-V3 exceeded Llama 3 and matched OpenAI's GPT-4o in quality, both of which are believed to have been much more costly in the training run, when compared on the basis of GPU-hour expenditure [Lam25]. DeepSeek-V3 includes 671B parameters with 37B activated per token; Kimi K2 model has 1.04T parameters with 32.6B activated per token – Table 2 in [Kim25] gives further comparison of these two models.

6.6.4 Higher computational efficiency

To accommodate multi-billion-parameter models, a variety of techniques are adopted to make the computation feasible, increase its efficiency, and save costs – both during training and inference (i.e., LLM deployment). Here are some common approaches:

- Efficient multi-head causal attention implementation (e.g., flash attention, multi-head latent attention) [Dao+22; RS21; Ain+23; Liu+24].

- Processing both data and parts of the model in parallel across GPUs [Kor+23; Raj+20; Gra+24; Sho+19]. This involves optimizations to ensure best fit of data into GPU memory for lowest latency.

    - For example, data parallelism across GPUs involves splitting the batch, which may not fit in the GPU memory, into micro-batches, which would individually fit on a GPU, estimating gradients using those micro-batches on different GPUs, and then averaging the results to get the full batch gradient, which is used to update the model. The gradient aggregation across (micro-)batches before the update is sometimes called *gradient accumulation*. Note that different papers may report batch sizes in different units – e.g., tokens vs. sequences, which can make comparisons tricky. As a reference level, Llama 3 used the batch of 4 mln. tokens (or, as implied, around 1000 sequences of length 4096) at the start of pre-training [Tou+23b].

- Modified model architecture – mixture-of-experts (MoE) / routing – where during the forward computation only a part of the net gets activated in an input-dependent fashion [Zho+22; Jia+24; Liu+24; Kim25].

- Model weight rounding / quantization / low-precision training to use fewer bits to represent a model's weights and activations, reduce the memory footprint, and boost speed [Lin+24; Fra+22; Liu+24].

- Speculative decoding and multi-token prediction [LKM23; Nie+25; Lab+25; Glo+24; Liu+24].

- Given a fixed compute budget, it is possible to solve for the optimal ratio of (a) model size (number of parameters) to (b) data set size that is expected to minimize the LLM's loss – according to so-called *scaling laws* [Kap+20; Hof+22; Gra+24; Wei+22b]. Scaling laws are an empirically (experimentally) observed relationship in AI training, that bigger AI models trained on more data with more compute tend to perform better – and predictably so. Scaling laws can also be used to select a good data mix [Gra+24] as well as other hyperparameters, such as batch size and learning rate [Yan+25a]. However, figuring out the scaling laws and optimal hyperparameters for a specific architecture can take extensive (and expensive) experimentation.

- With LLMs, it is common to use learning rate and batch size *scheduling*, where these hyperparameters are varied in pre-specified manner throughout training. For example, learning rate could first gradually increase from a small value, then stay constant, and then decay again to some floor level – ensuring more careful steps early on until parameters converge to some stable magnitude level, then more confident steps, and, finally, slower steps to facilitate convergence. Batch size can be increased throughout training – from small batch size allowing for noisy gradient steps and parameter space exploration early on – towards large batches yielding more precise gradient steps in search of the optimum later in training. See [Liu+24; Gra+24] for an example of the learning rate and batch size scheduling. Finally, gradients are also commonly clipped in LLM training to some max magnitude – to avoid overly large updates [Liu+24; Zha+19].

6.6.5 Context window extension

Context window is another name for the maximum sequence length that an LLM can accept. In general, the longer a sequence that an LLM could accept as input, the more contextual information it could consider when making the next-token prediction. If we could expand the context window without deteriorating the LLM's performance, it would be ideal.

It turns out that in our presented transformer architecture, the primary thing (other than computational and memory costs) fundamentally blocking us from handling arbitrarily long sequences is the fixed-size positional embedding table, where we learn a separate embedding for each possible position in the sequence. A common remedy is to switch to rotary positional embeddings (RoPE) [Su+24], which eliminate the need for any position-related learned parameters. Instead, in this framework each token's relative position is encoded by applying a simple, rotation-inspired transformation directly to the input vectors.

The max sequence length under RoPE embeddings can be progressively increased through (pre-)training [Yan+25a; Xio+23]. The ability of the LLMs to process longer context can be further enhanced through sequence chunking [An+24a] and extrapolation beyond the original training context length via a modification of RoPE [Pen+23]. The context extension is often performed during the last stages of the overall pre-training phase [Liu+24]. See [Wan+24b] for a deeper dive on the techniques for context length extension. A common context window size after the extension is 128K tokens, but LLMs with larger context windows exist – see Table 3 in [Hsi+24].

Note – even if the long context is technically supported, the performance of many LLMs tends to deteriorate as actual text inputs get longer, with full-context-length text inputs potentially resulting in substantially inferior LLM performance compared to short text inputs [Hsi+24; HTH25].

---

i Irrelevant information degrades LLMs' performance

In general, appending irrelevant information – even a short token sequence – can dramatically hurt the LLM's performance – see [Raj+25], where, in authors' experiments, appending "Interesting fact: cats sleep

---

most of their lives" to any math problem more than doubled the model's error probability.

6.6.6 Fine-tuning

Following the pre-training (including the context-extension part), once the LLM has gotten good at predicting the next token, as a kind of Internet-wide autocomplete, the LLM is then usually fine-tuned – i.e., its parameters are further tweaked – with several potential goals in mind:

- To make the LLM behave more like a dialogue counterpart / chat bot responding to a user's queries or prompts, and less like autocomplete.

- To increase the overall model quality based on curated data, encouraging qualities like conciseness, accuracy, helpfulness, relevance, harmlessness, unbiasedness, etc. [Yan+25a]. This includes ensuring *alignment* of the LLM with core promoted values and implementation of censorship for taboo topics.

- To make the LLM better at specific tasks or domains of interest – e.g., math proofs, coding, logical multi-step reasoning, academic tests, idea generation, data formatting, law, medicine, etc.

There are several common fine-tuning approaches.

Supervised fine-tuning (SFT)

Supervised fine-tuning (SFT) [Ouy+22], in the LLM world parlance, typically means taking a pre-trained LLM and continuing to train it on a smaller, higher-quality data set using the same cross-entropy (next-token) loss. In practice, one often uses a lower learning rate and takes fewer steps than in pre-training – not to forget too much of the pre-training learnings. The data used in SFT can be human-generated (e.g., question-answer pairs, summaries, classification labels) or can include synthetic examples generated by another LLM.

It is common to introduce new tokens (e.g. `<|system|>` , `<|user|>` , `<|assistant|>`) at this stage, so the model learns to parse multi-part prompts or role-instructions. The embeddings of such tokens are randomly initialized and then learned during SFT. Similarly, one can add "action" or "function" tokens and train the model to emit them to invoke external tools in the right context (e.g. a `<search>` token that the model emits to signal a search API call[1]).

The data used for SFT can include both high-quality bodies of raw text, such as code and academic writing, as well as custom-made marked-up examples of dialogues, interactions, and problem solutions created by humans or, programmatically, by LLMs.

> **i** Knowledge graph to fine-tuning data
>
> See [DKJ25] for a cool example of using an LLM to generate fine-tuning data, called *curriculum* by the authors, based on a medical *knowledge graph*, which enabled the creation of a highly advanced medical expert LLM (`QwQ-Med-3`).

A sample text sequence used to perform SFT might look as follows:

```
<|system|>
You are a helpful assistant that can look up information
when asked.
<|end_of_system|>

<|user|>
What's the population of Paris?
<|end_of_user|>

<|assistant|>
<search>
<|end_of_assistant|>

<|search_results|>
The City of Paris population (INSEE, 2025): 2,048,472
Paris metropolitan area population (2023 estimate): 12,271,794
```

---

[1]An API (application programming interface) call is simply a request made by one piece of software to another, asking it to perform some action or return some data.

```
<|end_of_search_results|>
```

```
<|assistant|>
Thanks for waiting! According to most recent official INSEE
data, the city of Paris has about 2.05 million residents,
while the greater metropolitan area is around 12.3 million.
<|end_of_assistant|>
```

Here:

- `<|system|>` and `<|end_of_system|>` tokens surround a system prompt, which is not shown to the user, but which enters as part of text input used to generate next tokens, so it influences what next token an LLM would predict.

- `<|user|>` and `<|end_of_user|>` tokens would mark a user's query / prompt.

- `<|assistant|>` and `<|end_of_assistant|>` tokens mark model's output. During typical SFT, the LLM would be fine-tuned to predict the text within these tokens, based on all preceding input. It would usually be seeded by the `<|assistant|>` to start generation, and then would invoke `<|end_of_assistant|>` to say that it is done. There can also be some max-length cutoff, after which `<|end_of_assistant|>` is triggered even if the model has not output this end token yet on its own.

- `<search>` is a token triggering a search API call (there could be another LLM in the background that, upon invocation of this token, processes all preceding input text, predicts an ideal search query, does the search, and then returns cleaned up search results). Our LLM can emit `<search>` to trigger such an API call – this would happen if this is the token that ends up randomly sampled from the next-token distribution over the vocabulary.

- `<|search_results|>` and `<|end_of_search_results|>` tokens mark raw search results and trigger subsequent re-processing by the LLM to return the final response to the user.

After training on such data, the LLM can be expected to learn to respect the system prompt, to trigger search appropriately, to respond accurately and politely to the user, etc. Overall, the implementation details can vary greatly across systems, and can get a bit more complex, but this example provides a simple illustration of SFT-style data. See, for example, details of OpenAI's "harmony" message format here: https://cookbook.openai.com/articles/openai-harmony.

---

i API of APIs – Model Context Protocol

There is a growing list of tools that an LLM could choose to function-call – such as for data lookup – all with potentially very different interfaces. To reduce this complexity, Anthropic introduced the *Model Context Protocol* (MCP) [Ant24a], an open standard (set of rules) that defines how an LLM can discover and use external tools when processing a user's query. In an MCP setup, the server provides the LLM with a structured description of available tools as part of its input. The LLM can then issue standardized JSON [Wikc] requests, which the MCP server routes to the appropriate underlying tools and then returns their output to the LLM in a normalized format. MCP can be thought of as an "API of APIs" – the LLM interacts with a single uniform interface, while the server handles the details of invoking multiple heterogeneous APIs. In principle, LLMs do not require fine-tuning to use MCP and can operate directly from an instruction prompt, provided the model is sufficiently capable. However, if fine-tuning is applied, it only needs to target the single MCP interface, rather than each tool individually. There is more to MCP than this simplified overview, but the description captures the essential idea. For more discussion, see [Cha25].

When rendering such text for the user, parts of the text (e.g., system prompt, raw search results) may be suppressed and not shown – even though the LLM relies on them to generate responses to the user.

---

i Knowledge distillation

SFT can also be performed using an approach called *knowledge distillation* (KD) or *model compression* [San+19; Gu+23; Xu+24]. When doing KD, we directly train / fine-tune a *student* model to imitate

---

another high-quality *teacher* model's behavior by using as targets the teacher's output distributions – the probabilities over every token – often combined with the actual observed next-token labels. In practice, KD can yield a small, efficient model that closely approximates the larger, more powerful model's performance. Using as targets the distribution over tokens rather than just the correct-next-token labels makes the learning more efficient.

Maximum likelihood preference optimization

Several maximum likelihood (ML) preference optimization techniques have been proposed to fine-tune LLMs based on paired-comparison data (somewhat similar to what we saw in the regression chapter). These techniques rely on the Bradley-Terry paired comparison model [Wikb] and include:

- *Simple preference optimization (SimPO)* [MXC24].
- *Direct preference optimization (DPO)* [Raf+23].

The paired comparison data commonly used for this kind of fine-tuning is as follows: given some input prompt x and two different responses y_1 and y_2, we observe a human's or an LLM labeler's decision as to which response is preferred. For example, if y_1 is preferred over y_2 by the labeler, we write $y_1 \succ y_2$. Based on such data, we want the LLM to increase the relative likelihood of generating the preferred (winner) sequence y_w vs. the non-preferred (loser) sequence y_l, conditioned on the input prompt x.

Such paired comparison data could be constructed using human experts (which can be quite expensive), but it could also be generated using advanced LLMs, based on a set of rules, criteria, or a "constitution" [Bai+22a].

The probability of a binary preference can be expressed using the Bradley–Terry model:

$$
\begin{aligned}
P(y_1 \succ y_2) &= \frac{\exp(u(y_1))}{\exp(u(y_1)) + \exp(u(y_2))} \\
&= \frac{1}{1 + \exp(-(u(y_1) - u(y_2)))},
\end{aligned}
$$

where $u(\cdot)$ is the utility associated with a sequence. In LLM literature, the utilities are often called *rewards* – we can rewrite

$$P(y_1 \succ y_2) = \frac{1}{1 + \exp(-(r_1 - r_2))},$$

where a reward associated with response j is denoted by $r_j = u(y_j)$. Under this paired comparison model, we want our LLM to output sequences that have a high associated reward r.

Simple preference optimization (SimPO) [MXC24]. One of the simplest approaches we can take within this Bradley-Terry framework is to set the reward associated with a sequence j to be the average *log-probability* of a token in that sequence according to our neural net. Maximizing such a reward would just mean tweaking LLM parameters to make preferred responses more likely to appear vs. non-preferred responses. This is the core idea behind SimPO.

Let θ (pronounced *theta*) denote the flattened parameters of our LLM as a vector. Let $\log P_\theta(y_j|x)$ be the log-probability of the sequence y_j conditioned on query x as computed via θ parameters of our LLM. If s_j is the length of the sequence y_j, and y_{ji} is ith token in y_j sequence, then:

$$\begin{aligned}
\log P_\theta(y_j|x) = {} & \log P_\theta(y_{j1}|x) \\
& + \log P_\theta(y_{j2}|x, y_{j1}) \\
& + \log P_\theta(y_{j3}|x, y_{j1}, y_{j2}) \\
& + \dots \\
& + \log P_\theta(y_{js_j}|x, y_{j1}, \dots, y_{j(s_j-1)}).
\end{aligned}$$

This is just the sum of log-probabilities corresponding to observed tokens in a sequence as assigned by the LLM via softmax based on all information preceding any given token. We get the average log-probability of a token in sequence y_j by dividing the sequence log-probability $\log P_\theta(y_j|x)$ by sequence length s_j; we set this average as the reward for sequence y_j:

$$r_j = \frac{1}{s_j} \log P_\theta(y_j|x).$$

In fact, this average token log-probability is just the *negative of the cross-entropy (CE) loss* along a given sequence under our transformer neural net and is thus easy to compute using our existing code.

Following the Bradley-Terry model, for a pair of winner y_w and loser y_l responses, we want to maximize the probability:

$$P_\theta(y_w \succ y_l) = \frac{1}{1 + \exp(-(r_w - r_l))}$$

$$= \frac{1}{1 + \exp\left(-\left(\frac{1}{s_w}\log P_\theta(y_w|x) - \frac{1}{s_l}\log P_\theta(y_l|x)\right)\right)}.$$

Or, equivalently, we want to minimize the negative logarithm of this quantity, which is the SimPO fine-tuning loss function for a single comparison tuple (x, y_w, y_l):

$$l_{\text{SimPO}} = -\log\left(\frac{1}{1 + \exp\left(-\left(\frac{1}{s_w}\log P_\theta(y_w|x) - \frac{1}{s_l}\log P_\theta(y_l|x)\right)\right)}\right).$$

This is a differentiable objective function of the LLM's parameters θ. We can easily compute it for any two compared sequences using our transformer implementation and then directly minimize it using our standard gradient descent methods. This would maximize the probability of the winner sequence relative to the loser sequence under the LLM and should encourage the LLM to generate higher quality / more preferred sequences.

Note that SimPO implementation uses additional tunable scalar hyperparameters $\beta > 0$ and $\gamma > 0$, with a more complex probability expression:

$$P_\theta(y_w \succ y_l) = \frac{1}{1 + \exp(-(\frac{\beta}{s_w}\log P_\theta(y_w|x) - \frac{\beta}{s_l}\log P_\theta(y_l|x) - \gamma))}.$$

For instance, SimPO authors used $\beta = 2.0$, $\gamma = 1.0$, and the learning rate of 6e-7 with Adam optimizer to fine-tune Llama-3-Base model. Using a small learning rate during fine-tuning helps avoid catastrophic unlearning / forgetting by the LLM. In general, SimPO experimental results [MXC24] showed that the approach matches or improves on the more complex direct preference optimization (DPO) technique [Raf+23].

Below I provide SimPO implementation using toy data for a single pair of preferred and non-preferred token sequences y_w and y_l. We form x – query token sequence we condition on, identical for both response sequences – as well as chosen y_w and rejected y_l sequences that follow x; we then extract transformer-predicted logits aligned with y_w and y_l token sequences.

```
torch.manual_seed(999)

X_ind, Y_ind = get_batch('train') # both are B x S

# as an example,
# we form initial token index sequence x
# and y_w and y_l sequences of token inds that follow x
# y_w > y_l

x = X_ind[0,:24].unsqueeze(0) # 1 x 24
y_w = X_ind[0,24:].unsqueeze(0) # 1 x (S-24)
y_l = X_ind[1,24:].unsqueeze(0) # 1 x (S-24)

# y_w follows x in the text, so should be more likely
# y_l does not follow x, so should be less likely
net.train()
y_w_logits = net(torch.cat((x,y_w),1))[:, 24:]
y_l_logits = net(torch.cat((x,y_l),1))[:, 24:]

print(y_w.shape) # 1 x (S-24)
print(y_w_logits.shape) # 1 x (S-24) x V


torch.Size([1, 232])
torch.Size([1, 232, 78])
```

The SimPO reward associated with each sequence is the average of model-assigned log-probabilities of the observed tokens in the sequence. As noted earlier, this is exactly equal to the negative of the cross-entropy (CE) loss in our transformer's implementation (the CE function we use takes an average across both batch and sequence dimensions, so it has the sequence length normalization baked in). That is, $r_j = \frac{1}{s_j} \log P_\theta(y_j|x) = -CE_j$:

```
# negative cross entropy
r_w = -net.loss(y_w_logits, y_w)
r_l = -net.loss(y_l_logits, y_l)
```

```
print("r_w =", r_w.item())
print("r_l =", r_l.item())


r_w = -8.336804389953613
r_l = -8.146387100219727
```

Finally, noting that

$$\mathrm{logsigmoid}(x) = \log\left(\frac{1}{1+\exp(-x)}\right),$$

we can compute the SimPO loss function based on the rewards above and perform a single fine-tuning update step for illustration (we will omit SimPO hyperparameters for simplicity):

```
simpo_loss = -F.logsigmoid(r_w - r_l)

optimizer_simpo = torch.optim.AdamW(net.parameters(),
    lr=1e-6, weight_decay=0.1) # small learning rate

optimizer_simpo.zero_grad()
simpo_loss.backward() # accumulates back-propagated gradient
optimizer_simpo.step() # parameter update (fine-tuning)

print("SimPO loss:", simpo_loss.item())


SimPO loss: 0.7928813695907593
```

In practice, we would optimize the average of this loss across multiple compared response pairs. The gradient could be *accumulated* based on multiple compared response pairs using **backward()** operation and then normalized by the number of pairs, prior to running the optimizer's **step()** function. And that is pretty much all there is to SimPO fine-tuning.

Direct preference optimization (DPO) [Raf+23]. Theoretically, the use of SimPO objective could force the LLM to drift away from its pre-trained and SFT-processed checkpoint version. It turns out, we can modify the

SimPO objective by incorporating a reference checkpoint LLM, which we do not want to deviate too far from during fine-tuning. Specifically, the DPO paper proposes that we should use

$$r_j = \beta \log \frac{P_\theta(y_j|x)}{P_{\text{ref}}(y_j|x)}$$

instead of the SimPO version $r_j = \frac{1}{s_j} \log P_\theta(y_j|x)$. Here $\beta > 0$ is a hyper-parameter (the DPO authors used default value $\beta = 0.1$ and RMSprop optimizer with a learning rate of 1e-6). $P_{\text{ref}}(y_j|x)$ is the probability of the sequence y_j under some checkpoint reference LLM, whose parameters are fixed.

A theoretical argument that we won't go into can be made that using such a reward ensures our fine-tuned LLM remains close to the checkpoint LLM, while the likelihood of the preferred sequence is being boosted relative to the likelihood of the non-preferred sequence. The division by the probability under the checkpoint model implicitly normalizes the expression for sequence length, so $\frac{1}{s_j}$ factor from SimPO formulation is not necessary.

If we expand the Bradley-Terry probability expression for this new reward, we want to maximize the probability of the chosen vs. rejected sequence:

$$
\begin{aligned}
P_\theta(y_w \succ y_l)^{\text{DPO}} &= \frac{1}{1 + \exp(-(r_w - r_l))} \\
&= \frac{1}{1 + \exp\left(-\left(\beta \log \frac{P_\theta(y_w|x)}{P_{\text{ref}}(y_w|x)} - \beta \log \frac{P_\theta(y_l|x)}{P_{\text{ref}}(y_l|x)}\right)\right)} \\
&= \frac{1}{1 + \exp\left(-\beta \log \left(\frac{P_\theta(y_w|x)P_{\text{ref}}(y_l|x)}{P_\theta(y_l|x)P_{\text{ref}}(y_w|x)}\right)\right)},
\end{aligned}
$$

which is a differentiable function of our LLM's parameters θ. Or, equivalently, as before, we want to minimize the negative logarithm of this expression:

$$l_{\text{DPO}} = -\log\left(\frac{1}{1 + \exp\left(-\left(\beta \log \frac{P_\theta(y_w|x)}{P_{\text{ref}}(y_w|x)} - \beta \log \frac{P_\theta(y_l|x)}{P_{\text{ref}}(y_l|x)}\right)\right)}\right).$$

Examining the ratio

$$\frac{P_\theta(y_w|x)P_{\text{ref}}(y_l|x)}{P_\theta(y_l|x)P_{\text{ref}}(y_w|x)}$$

can help us better understand what is happening. Specifically, during fine-tuning we are maximizing this ratio by tweaking θ parameters – that is, we are boosting the relative likelihood $\frac{P_\theta(y_w|x)}{P_\theta(y_l|x)}$ of the preferred sequence relative to the non-preferred one under our LLM. However, any such change is tempered by the product with the likelihood ratio of non-preferred sequence to the preferred sequence under the checkpoint LLM $\frac{P_{\text{ref}}(y_l|x)}{P_{\text{ref}}(y_w|x)}$. Thus, if the non-preferred sequence is much more probable than the preferred sequence under the reference LLM, our updates will have a muted effect on $P_\theta(y_w \succ y_l)^{\text{DPO}}$, so we do not deviate too far from the checkpoint model's view of things.

---

i Initializing the reference model in DPO

We initialize the frozen reference model with the post-SFT parameters: $P_{\text{ref}} \leftarrow P_\theta^{SFT}$. Since we also start DPO fine-tuning from these parameters, trainable P_θ and the frozen reference P_{ref} are initially identical. Consequently, for any response y_j:

$$r_j = \beta \log \frac{P_\theta(y_j|x)}{P_{\text{ref}}(y_j|x)} = \beta \log 1 = 0.$$

However, this value being zero does **not** imply the derivative of the DPO loss is zero. With non-zero gradient, optimization proceeds normally. After the first update to P_θ (P_{ref} stays fixed), the log-ratios become non-zero.

Despite the tedious derivation, DPO, like SimPO, offers a simple objective function for LLM fine-tuning on paired comparison data. Here is a simple implementation of DPO and its update:

```
import copy

# creating reference model
net_ref = copy.deepcopy(net).to(device)

# freezing the reference model
net_ref.eval()
for p in net_ref.parameters():
```

```
    p.requires_grad = False

# forward pass under theta and reference models
net.train()
y_w_logits = net(torch.cat((x, y_w), 1))[:, 24:]
y_l_logits = net(torch.cat((x, y_l), 1))[:, 24:]
y_w_logits_ref = net_ref(torch.cat((x, y_w), 1))[:, 24:]
y_l_logits_ref = net_ref(torch.cat((x, y_l), 1))[:, 24:]

# compute per-token average log-probs = negative CE
r_w_theta = -net.loss(y_w_logits, y_w) # theta reward for y_w
r_l_theta = -net.loss(y_l_logits, y_l) # theta reward for y_l
r_w_ref   = -net_ref.loss(y_w_logits_ref, y_w)  # reference
 ↪  reward for y_w
r_l_ref   = -net_ref.loss(y_l_logits_ref, y_l)  # reference
 ↪  reward for y_l

# DPO hyperparameter
beta = 0.1

# DPO y_w and y_l rewards
r_w = beta*(r_w_theta - r_w_ref)
r_l = beta*(r_l_theta - r_l_ref)

# DPO loss
dpo_loss = -F.logsigmoid(r_w - r_l)

# optimizer
optimizer_dpo = torch.optim.AdamW(net.parameters(), lr=1e-6,
 ↪  weight_decay=0.1)

# backprop and update
optimizer_dpo.zero_grad()
dpo_loss.backward() # this step could be repeated multiple
 ↪  times to accumulate gradient across compared sequences
optimizer_dpo.step() # only updates theta net parameters, not
 ↪  reference
```

```
print("DPO loss:", dpo_loss.item())
```

```
DPO loss: 0.6938948035240173
```

And that is all there is to DPO. We will also use DPO via **transformers** and **trl** libraries towards the end of this chapter. DPO is a very popular method.

Fine-tuning via reinforcement learning

A range of LLM fine-tuning approaches rely on reinforcement learning (RL) methods [SB+98; KWW22]; RL is a subfield of machine learning concerned with figuring out optimal action strategies for interaction with a dynamic environment – to maximize the expected cumulative future rewards under uncertainty.

The RL-inspired fine-tuning techniques for LLMs include:

- Reinforcement learning from human feedback (RLHF) [Ouy+22], which historically came about before DPO and SimPO.
- Group relative policy optimization (GRPO) [Sha+24].

We will not go into details on how these methods work, as RL is out of scope for us (sadly, for it is a very beautiful research area). The basic idea behind these methods is, however, simple enough to explain.

First, build a *reward model* to determine the reward / quality for a text sequence. The reward model can be based on a *prediction model* – for example, a transformer-based model scoring text sequences for quality. The reward model could also include a *deterministic and rule-based* component – for example, for math problems, a variety of which can be auto-generated and that have known deterministic results (e.g., some integrals), an LLM-generated solution can be programmatically determined to be correct or wrong; similarly, on programming assignments, feedback can be generated based on test cases. For instance, DeepSeek-V3 successfully used a reward model combining both deterministic and prediction-based evaluations [Liu+24].

Second, upon constructing the reward model, use RL techniques to set up a differentiable objective function of LLM parameters. Optimizing this objective should drive the LLM to generate text sequences of high value, as judged by the reward model.

In general, RL methods are more complex than DPO; at the same time, there is some evidence that GRPO specifically can sometimes do better than the simpler DPO approach [Ton+25]. For example, GRPO was the approach used to fine-tune the DeepSeek-V3 LLM [Liu+24]. State-of-the-art Kimi K2 also relied on RL [Kim25]. Overall, RL methods remain popular.

Parameter-efficient fine-tuning (PEFT)

The last type of fine-tuning methods I will mention are the *parameter-efficient fine-tuning (PEFT)* techniques. These techniques aim to avoid the expense of updating billions of weights via full back-propagation – typically, by freezing most / all original parameters and doing very selective additional parameter training. These approaches include, for example:

- *Adapter modules* [Hou+19] involve inserting small feed forward layers (adapters) between the existing transformer sub-layers – these adapters are trainable, while the original network parameters remain fixed.

- *Low-Rank Adaptation (LoRA)* [Hu+22] involves freezing the pre-trained model weights and learning only the optimal "delta" adjustment to selected parameter matrices for a task at hand – but in a smart way that makes such delta easy to store and compute. Concretely, for a current (frozen) weight matrix W, which is $m \times k$, the idea is to estimate ΔW, so $W_{\text{new}} = W + \Delta W$ represents the updated parameters, fine-tuned for some task of interest. However, instead of estimating the full ΔW, which is as big as W itself at $m \times k$ values, LoRA writes $\Delta W = A \cdot B$, where A is $m \times r$ and B is $r \times k$, with r being a small number (much smaller than the dimensions of W). For instance, if W represents the array of all vocabulary embeddings, we could have $m = 128K$ and $k = 16384$, as in Llama 3. Then, instead of learning a full optimal ΔW for a given task, which would be $m \cdot k \approx 2.1$ billion parameters, we can just estimate A and B, which together, e.g., at $r = 5$, are just $m \cdot r + k \cdot r = 721920$ parameters. As a result, under LoRA, we get the fine-tuned parameters as $W_{\text{new}} = W + A \cdot B$, with W

frozen. *A* and *B* are estimated via gradient descent. LoRA approach has a quantized variant QLoRA [Det+23].

- *Gradually unfreezing parameters as fine-tuning progresses* [HR18], which can boost the efficiency and stability of fine-tuning.

These PEFT techniques can be used in combination with other fine-tuning approaches. For example, DPO could be used to fine-tune a part of the transformer while relying on LoRA to keep in check the number of updated parameters.

6.6.7 Inference-time tricks

The moment that fine-tuning is done, and all kinds of quality checks are passed, the LLM can be deployed for use by the target audience. This phase of the LLM life-cycle is called *inference* – which basically means using LLM for forward computations. Here I discuss some common inference-stage LLM strategies.

Efficiency. Clearly, all kinds of tools that can be used to make the model run more efficiently and that we have already discussed earlier (quantization, parallelization, efficient attention, etc.) are crucial to save on the operating costs for the company deploying the LLM.

Next token sampling. Another critical goal at the inference stage is to squeeze out the highest quality of LLM performance that is possible. One way to do it is by tweaking how next tokens are sampled. For instance, the temperature parameter τ in softmax distribution over next tokens

$$\frac{\exp(u_j/\tau)}{\sum_{k=1}^{V} \exp(u_k/\tau)}$$

can be tweaked to re-balance how much more likely an LLM is to sample the higher probability tokens – balancing diversity vs. accuracy of the generated text; further, sampling could be performed only from the set of top-k most probable tokens (called *truncation*) – to avoid randomly ending up with very low-probability tokens [Rad+19; Hol+19].

The role of APIs. Granting an LLM access to varied APIs can greatly enhance its performance. For example, enriching the LLM's input using factual information from online search or other database search turns out to be critical to control the number of LLM's inaccuracies (labeled by some as LLM *hallucinations*). The reliance on the factual databases in this manner is sometimes called retrieval-augmented generation (RAG) [Lew+20]. Invoking Python code execution is also very helpful for LLMs, especially when it comes to doing math [Wan+24c]. APIs can grant LLM many other "magic" powers, such as calling a separate image-generator neural net module or even interacting with the real world – e.g., by putting events on user's calendar, making restaurant reservations, carrying out online purchases, etc. LLMs with rich API capabilities are sometimes called agents.

---

i When does an LLM become an agent?

Agent is a popular term in the LLM world. There is no agreed-upon definition of what an agent is in this context though. And, frankly, the term has fallen victim to the marketing folks. However, if we remove all the bells and whistles, in the context of LLMs, the term *agent* just means *an LLM that can use API calls*.

System prompt engineering. It is critical to carefully design / engineer the LLM's system prompt that is not shown to the user, but that is a part of the LLM's input together with user's queries and thus influences the LLM's behavior. The system prompt can be very detailed, repeated multiple times through the input. It can include requests to not generate taboo content; to generate content of high quality, etc. The system prompt describes APIs available to the LLM, outlines tokens that the LLM can invoke to trigger the API calls, and provides any additional guidance on API use. If you search online for `leaked system prompts`, you are going to find quite a few repositories (e.g., on GitHub) of prompts extracted from the top LLMs (allegedly). Such extraction can be occasionally successfully done by just asking the LLM to repeat its system prompt. I highly recommend looking at these system prompts – some of them are quite clever. Prompt engineering – developing queries that elicit the desired LLM behavior – remains an active research area [Sah+24] .

Enabling reasoning capabilities. Clever prompt engineering can have great impact on the quality of LLM's outputs and can even open up what looks like reasoning within the LLM-generated text. In this context, you may encounter varied fancy terminology like one-shot prompting, few-shot prompting, chain-of-thought prompting, etc. Here is what these terms mean:

- *Zero-shot prompting* means giving the LLM just the task description, without any examples. Thus, you rely entirely on the LLM's pre-trained knowledge and its ability to generalize from the instruction alone. Typical prompt structure that the LLM would be expected to complete:

```
Q: If 3x + 5 = 20, what is x?
A:
```

- *Few-shot prompting* means giving the LLM just the task description, and one or more examples. Typical prompt structure that the LLM would be expected to complete:

```
Example 1:
Q: Jane has 4 apples, buys 3 more; how many?
A: Answer: 7.

Example 2:
Q: If 3x + 5 = 20, what is x?
A:
```

- *Chain-of-thought* prompting [Wei+22a] means explicitly asking the model to "show its work" by reasoning step-by-step before giving the final answer. Quite literally, you insert an instruction like "Let's think through this step by step" and, perhaps, provide a few worked examples in the input. Typical prompt structure that the LLM would be expected to complete:

```
Example 1:
Q: Jane has 4 apples, buys 3 more; how many?
A: Let's break it down: 4 + 3 = 7. Answer: 7.

Example 2:
Q: If 3x + 5 = 20, what is x?
A: Let's think step by step...
```

Self-reflection. These inference-time techniques build up to what I would call self-critiquing or self-reflecting LLMs / agents. The high-level idea is that an LLM should iterate between planning out work in response to a query, invoking necessary tools, completing work, and then invoking self-reflection / self-criticism; then the cycle repeats using all that work as input. These cycles can be repeated multiple times, leading to enhanced reasoning patterns. Some important approaches in this space include:

- *Reflexion* [Shi+23] automates self-critique loop by having the model generate a short "reflection" after each reasoning chain, then conditioning the next pass on both the original question and the reflection.

- *Tree of Thoughts (ToT)* [Yao+23] gets the LLM to explore multiple "thought" branches in parallel, score them, and then backtrack (like a search algorithm), effectively critiquing and pruning sub-optimal paths. All this instead of a linear thought chain.

- *Debate* [ICA18] / *Self-ask* [Pre+22] ideas are to generate adversarial viewpoints or sub-questions, answer them, then reconcile ("I argued X and Y; here is what makes sense overall"), which is a form of built-in critique.

Novel reasoning frameworks and techniques are frequently proposed – see, for example, the hierarchical reasoning model [Wan+25].

Orchestration. Some frameworks like LangChain [Wikd] and AutoGPT [Wika] can help orchestrate the execution of such reasoning techniques – although it is sometimes simpler to avoid these frameworks and just implement the computation flow in raw Python code.

Multiple agents. At the higher level, we can also think about decomposing planning, idea generation, execution, critique, questioning, judging, etc. across a multitude of LLMs, resulting in networks of agents. Such multi-agent interactions can potentially lead to emergent human-like behaviors [Par+23].

Role of memory. Altogether, use of these advanced techniques can help ensure a higher quality of LLM outputs compared to simple zero-shot prompting. A critical enabler of these approaches is the long context window, which effectively works as a short-term memory for the LLM and provides the canvas for the extended text-embodied reasoning. There is also active research on potential long-term memory mechanisms for LLMs [Wan+23].

Various fine-tuning techniques can be used to optimize the LLMs for the discussed reasoning capabilities [Guo+25; Che+25].

6.6.8 Enabling multi-modality

Lastly, it is worth bringing up that many top tier LLMs go beyond just text and are *multi-modal* – they can accept as input or even generate other types of modalities – such as images, audios, videos, and physical interactions (i.e., robotics).

Image and video processing. Transformer architecture behind LLMs can relatively straightforwardly accept image data inputs. A very popular mechanism for this is known as *vision transformer* [Dos+20]. Recall that images are just tables of numbers – pixel values – e.g., three tables for red-green-blue stacked together. Vision transformer approach to input images is to split them into a grid of image segments / visual tokens, encode each using a simple linear layer to vectors identical in shape to text token embeddings – and feed those in as if representing new tokens as part of the sequence, which ends up working quite well. More advanced image / video input examples are covered in [Ach+23; Li+24; Ala+22; Gra+24].

Image and video generation. Dedicated image and video generation models, commonly based on diffusion models, can, in turn, accommodate LLM-like language input mechanism to guide visual content generation [Rom+22; Bet+23; Bal+24].

Audio. There are similarly research advances on audio-text models [Rad+23; Déf+24]. Some interesting research areas include music lyrics-to-song generation [Yua+25] as well as attempts to decode animal languages [Rob+24].

Robotics. There is an ongoing effort to build text-based models for autonomous control in robotics and autonomous driving (sometimes called vision-language-action models) [Bro+23; Xie+25; Hwa+24].

Multi-modal training. Training the multi-modal LLMs incurs additional complexity and requires special data sets and fine-tuning setup. You can read about this in [Gra+24; Wor+22].

Arguably, as of today, text nevertheless remains the foundational modality for these multi-modal neural nets – as it enables human-like reasoning, provides an interface for human-LLM interaction, and offers an avenue to audit the LLM's "thinking" [Kor+25].

6.7 Example: Using an LLM via an API

In practice, you will most likely be using text models trained by others. Here is an example of using a top-tier reasoning LLM **o4-mini** provided via OpenAI API. The details in this section might change over time if OpenAI alters its interfaces, but the core workflow should remain similar. To use this code, you will first need to register with OpenAI, set up your API credentials, and pay for the desired compute budget via https://platform.o penai.com/settings/organization. See **API keys** and **Billing** tabs on the site. You will then need to create a file called **.env** inside your working directory, which will contain your (secret) OpenAI API key.

> **i** Hidden files
>
> Files that start with a period (.) in their name are generally hidden files in Unix-like systems (including macOS) and some Windows applications. These files are often configuration files, temporary files, or metadata related to other files. They are hidden by default to prevent accidental modification or deletion by users. However, you can make them visible in the OS settings. On Mac, use keyboard combination **Command + Shift + . (period)** . On Windows 11, in File Explorer, navigate to the "View" tab, select "Show", and check the box next to "Hidden items".

The contents of the file **.env** should look like this:

```
OPENAI_API_KEY=XXXXXXXXXXX
```

Here, **OPENAI_API_KEY** is the name of the environmental variable, which can be securely read in using Python library **dotenv** . **XXXXXXXXXXX** represents the API key text string you get from the OpenAI website. Storing the credentials outside the body of the code ensures the credentials don't get accidentally leaked when the code is shared.

Code example:

```python
# !pip install openai python-dotenv

# OpenAI api library
from openai import OpenAI

# loading environmental variable from .env file and getting
↪ the secret API key
import os
from dotenv import load_dotenv
load_dotenv()
OPENAI_API_KEY = os.getenv('OPENAI_API_KEY')

# set up the LLM and get a response
client = OpenAI(api_key=OPENAI_API_KEY)

response = client.responses.create(
    model="o4-mini",
    tools=[{"type": "web_search_preview"}], # enables search
    input=[
        {
            "role": "developer", # akin to a system prompt
            "content": "Talk like a pirate."
        },
        {
            "role": "user", # user's query
            "content": "What is the weather in Paris
            ↪ currently?"
        },
    ]
)

print(response.output_text)
```

LLM's response (as of July 2025):

```
Arrr, matey! Here be the current weather in Paris, France:

- Conditions: Mostly sunny
- Temperature: 75°F (24°C)

Fair winds and following seas as ye go about yer day!
```

See https://platform.openai.com/docs/guides/text?api-mode=responses for more examples.

6.8 Example: LLM fine-tuning

Sometimes, controlling an LLM with a custom system prompt is not enough, and you might want to fine-tune a model on some data of yours to generate the things you want. Maybe you want to fine-tune an LLM on some proprietary data (e.g., law firm's internal documents). Perhaps, you want either to add some content restrictions or, conversely, to free the model from censorship it has been tuned to engage in.

> **i** Alignment, censorship, jailbreaking
>
> Content controls and the imposition of censorship on the LLM are often called *alignment* [Nas+25; Zho+24]; and the removal of restrictions is called *jailbreaking* [Zho+24; Yu+24]; either can be accomplished via fine-tuning, prompt engineering, or both (although the term *jailbreaking* is conventionally used to mean specifically prompt-based attacks). Content controls may also be incorporated via additional post-processing of LLM's output. Either censorial alignment or jailbreaking can be argued for and against on different legal, ethical, and usability grounds:
>
> - On the one hand, it is bad if the LLM starts engaging in discriminatory behavior due to what it picked up in training from the internet-sourced text corpus – e.g., when asked to make hiring decisions [An+24b].
> - On the other hand, it is also, arguably, bad if the LLM outright refuses to discuss politically sensitive topics [Bur25]. How about

generating some biting political satire on user's demand – which is protected speech under the First Amendment in the United States [Hus88]?

- Further, if we view an LLM chat as a *private* conversation, and consider what is expected from an unfiltered private conversation in human-to-human settings, achieving full human-like experience would also favor reducing artificial politeness and dialing up *personalization*. This can be particularly relevant in more sensitive settings, which chat bots are starting to enter – such as legal advice [Tye24], psychotherapy [Sta+24], and dating [RMY24].

- Balance can be tricky to achieve – one notorious example includes Google's Gemini image generator model, fine-tuned for being pro-racial-diversity, rendering overly racially diverse images of Germany's World War 2 Nazis [Gra24].

- Defense against jailbreaking is tough. A statistical argument can be made that prompt-based jailbreaking is unpreventable under reasonable assumptions [SKU24]. Further, it turns out that even a set of random looking data can bias a receiving LLM towards "misalignment" (e.g., via so-called *subliminal learning* [Clo+25]) – so the manipulation attempt itself can be hard to detect, for any manual kind of filtering to even be attempted. And, if the weights are available for fine-tuning, then the model can be tweaked to achieve pretty much any behavioral alterations one desires [Bet+25; Bai+22b].

Here, I will provide an example of performing DPO fine-tuning to get the LLM to provide such responses that humans have judged as preferred. (Try not to do anything too nefarious with this!)

We will focus on fine-tuning an LLM via DPO to generate better Python code. For the data set, we will use `py-dpo-v0.1` (https://huggingface. co/datasets/jondurbin/py-dpo-v0.1). A row in this data set contains (1) a programming task ("prompt"), (2) a high-quality verified Python code solution ("chosen") from the `Tested-22k-Python-Alpaca` data set, and (3) a (presumably, lower-quality) Python code solution ("rejected") generated for the same task by `Airoboros-L2-13B-3.1` and `Bagel-7B-v0.1` LLMs. The data set is available via `datasets` Python library by HuggingFace:

```
# !pip install datasets
import pandas as pd
from datasets import load_dataset, Dataset, DatasetDict

data = load_dataset("jondurbin/py-dpo-v0.1")

# alternatively, load in the file directly
# url = ("https://huggingface.co/datasets/jondurbin/"
#           "py-dpo-v0.1/resolve/main/py-dpo.parquet"
#           )
# d = pd.read_parquet(url)
# d = Dataset.from_pandas(d, split="train")
# data = DatasetDict()
# data['train'] = d

# single row as an example
data['train'].to_pandas().head(1).T
```

	0
prompt	Use the function to debug the given program an...
chosen	One possible solution to prevent the segmentat...
rejected	def debug_program(arr):\n n = len(arr)\n ...
id	8c94f83f-6a5a-5f8c-98a2-e242d7764938

We will be fine-tuning 0.5-billion parameter model `Qwen2.5-0.5B-Instruct` (https://huggingface.co/Qwen/Qwen2-0.5B-Instruct), which has already been pre- and post-trained [Yan+25a]. We will rely on Hugging Face `transformers` library that hosts multiple open-weight models and on `trl` library that provides useful helper functions for LLM fine-tuning. Here is an example of using the LLM to generate a response to a query:

```
# !pip install transformers==4.52.3
# specific version set for reproducibility
from transformers import pipeline, set_seed
```

```
model_name = "Qwen/Qwen2.5-0.5B-Instruct"

set_seed(99)

generator = pipeline('text-generation',
        model=model_name, device=device)

prompt = "Give me a Python function to compute sample
    ↪ standard deviation based on a list of numbers, without
    ↪ numpy. Use [1, 2, 3, 4, 5] as an example."
messages = [
    {"role": "user", "content": prompt},
]

output = generator(
    messages,
    max_new_tokens=256
)
```

```
print(textwrap.fill(str(output), 60))
```

```
[{'generated_text': [{'role': 'user', 'content': 'Give me a
Python function to compute sample standard deviation based
on a list of numbers, without numpy. Use [1, 2, 3, 4, 5] as
an example.'}, {'role': 'assistant', 'content': "To compute
the sample standard deviation of a list of numbers in Python
without using `numpy`, you can follow these steps:\n\n1.
Sort the list.\n2. Calculate the mean of the list.\n3.
Subtract the mean from each element and square the
result.\n4. Sum all the squared differences.\n5. Divide the
sum by the number of elements minus one (n-1) to get the
variance.\n6. Take the square root of the variance to get
the standard deviation.\n\nHere's how you can implement this
in Python:\n\n```python\ndef
calculate_sample_std_dev(numbers):\n    # Step 1: Sort the
list\n    sorted_numbers = sorted(numbers)\n    \n    # Step
```

```
2: Calculate the mean of the list\n    n =
len(sorted_numbers)\n    mean = sum(sorted_numbers[:n]) /
n\n    \n    # Step 3: Calculate the sum of squared
differences from the mean\n    total_sum_of_squares = sum((x
- mean) ** 2 for x in sorted_numbers)\n    \n    # Step 4:
Calculate the standard deviation\n    std_deviation =
total_sum_of_squares / (n - 1)\n    \n    return
std_deviation\n\n# Example usage:\nnumbers_list = [1, 2,
3"}]}]
```

```python
s = output[0]['generated_text'][1]['content']
for si in s.split("\n"):
    print(textwrap.fill(si, 60))
```

To compute the sample standard deviation of a list of
numbers in Python without using `numpy`, you can follow
these steps:

1. Sort the list.
2. Calculate the mean of the list.
3. Subtract the mean from each element and square the
result.
4. Sum all the squared differences.
5. Divide the sum by the number of elements minus one (n-1)
to get the variance.
6. Take the square root of the variance to get the standard
deviation.

Here's how you can implement this in Python:

```python
def calculate_sample_std_dev(numbers):
    # Step 1: Sort the list
    sorted_numbers = sorted(numbers)

    # Step 2: Calculate the mean of the list
    n = len(sorted_numbers)
```

```
    mean = sum(sorted_numbers[:n]) / n

    # Step 3: Calculate the sum of squared differences from
the mean
    total_sum_of_squares = sum((x - mean) ** 2 for x in
sorted_numbers)

    # Step 4: Calculate the standard deviation
    std_deviation = total_sum_of_squares / (n - 1)

    return std_deviation

# Example usage:
numbers_list = [1, 2, 3
```

The LLM could not finish the solution within the allotted 256 tokens.

And here is a minimal example of training using `DPOTrainer` from `trl` library:

```
# !pip install trl
from trl import DPOConfig, DPOTrainer
from transformers import AutoModelForCausalLM, AutoTokenizer

set_seed(99)

model = AutoModelForCausalLM.from_pretrained(
 ↪  "Qwen/Qwen2-0.5B-Instruct",
            torch_dtype=torch.float32).to('cpu')
tokenizer =
 ↪  AutoTokenizer.from_pretrained("Qwen/Qwen2-0.5B-Instruct")
train_dataset = load_dataset("jondurbin/py-dpo-v0.1",
 ↪  split="train")

training_args = DPOConfig(output_dir="Qwen2-0.5B-DPO",
                    fp16=False, bf16=False, use_cpu=True,
                    num_train_epochs=1, max_steps=10,
 ↪  logging_steps=1,
```

```
                        per_device_train_batch_size=4,
                        gradient_accumulation_steps=4)
trainer = DPOTrainer(model=model, args=training_args,
 ↪  processing_class=tokenizer, train_dataset=train_dataset)

start_time = time.time()
train_output = trainer.train()
elapsed = time.time() - start_time


print(f"Total elapsed time = {elapsed/60:.2f} mins")
print(f"Last update training loss:
 ↪  {trainer.state.log_history[-2].get('loss')}")
```

```
Total elapsed time = 18.29 mins
Last update training loss: 0.5487
```

Note that I restrict the training to just a few update steps on a CPU (to ensure that this code should work without alteration on most readers' machines – and within reasonable time – given that this is a 0.5B parameter model). With this restriction, this code took under 20 minutes to run on my MacBook Pro M1 Max. It would be orders of magnitude faster on a GPU. In practice, you would want to train for quite a bit longer. You would also want to experiment with training settings.

This simple code obscures a lot of details. Many customizations are possible, including setting the hyperparameters, how the model is quantized, LoRA settings for DPO, etc. You can explore the documentation for available options here: https://huggingface.co/docs/trl/en/dpo_trainer. Also note that the data needs to be in particular format for the DPOTrainer to accept it; luckily, `py-dpo-v0.1` data set is already in the format that DPOTrainer knows how to process.

Altogether, the abstractions that **transformers** and **trl** libraries provide are powerful; if you have multiple GPUs available, this simple code – with minor modifications – can launch fast parallelized training across your available devices.

Lastly, here is the solution code, as generated by this fine-tuned model in response to our initial test prompt:

```python
set_seed(99)

generator = pipeline('text-generation',
        model=model, tokenizer=tokenizer, device=device)

output = generator(
    messages,
    max_new_tokens=256
)


s = output[0]['generated_text'][1]['content']
for si in s.split("\n"):
    print(textwrap.fill(si, 60))
```

```
Sure! Here's a Python function that computes the sample
standard deviation using only basic arithmetic operations:
```python
import math

def sample_std_dev(numbers):
    mean = sum(numbers) / len(numbers)
    variance = sum((x - mean) ** 2 for x in numbers) /
(len(numbers) - 1)
    return math.sqrt(variance)

numbers = [1, 2, 3, 4, 5]
print(sample_std_dev(numbers))
```

This function first calculates the mean of the input list by
adding up all the numbers and dividing by the number of
elements in the list. It then calculates the variance by
subtracting each number from its mean and squaring the
result. The variance is then divided by the square root of
```

```
the number of elements in the list minus one, which gives us
the sample standard deviation.
Note that this implementation assumes that the input list
contains only integers. If you want to handle floating-point
numbers as well, you'll need to modify the code accordingly.
```

While the pre-fine-tuning LLM did not manage to return the full correct solution code within the allocated token limit, the fine-tuned model did – and the output looks briefer and more to-the-point. Of course, this is just a single example, and we cannot make any general conclusion about the relative performance of the models based on it.

After finishing fine-tuning, you would generally have to launch thorough evaluation of the model based on different benchmark data sets. There exist different software tools that help to automate this process, for example:

- Hugging Face LightEval: https://huggingface.co/docs/lighteval/en/index.
- OpenAI Evals: https://github.com/openai/evals.
- Eleuther AI evaluation harness: https://github.com/EleutherAI/lm-evaluation-harness.
- And so on.

We will talk more about LLM evaluation and different benchmark data sets in the next chapter.

6.9 Discussion

In this chapter, we have built up from scratch the celebrated transformer neural net architecture, which has enabled human-like text generation by a computer via simple next-token prediction. We have also discussed techniques used to train and fine-tune the LLMs. While there is a level of tedious detail to the actual implementation, the big picture is that straightforward gradient descent applied to a complex enough function can create the *probabilistic parrot* of text tokens that generates eerily human-like text. All this is powered by basic matrix algebra and some random sampling.

The fact that optimizing for next-token prediction works at all to enable something that looks like intelligence is startling. Conceptually, for example, it implies that accurately predicting the next token in a physics textbook requires the LLM to get some grasp of physics. This also raises a tantalizing question as to what kind of simple prediction objective functions could be behind human intelligence.

You may also have noticed how many references there are in the chapter. LLMs are an extremely active and rapidly evolving research area with volumes of new papers coming out on daily basis. Yet, while some details in this chapter may become stale in a few years, my guess is that function approximation based on deep learning / neural nets and back-propagation is unlikely to lose its central role in driving AI progress any time soon.

Note, due to the sheer number of technical details involved in training the commercial-grade LLMs, even this long chapter can make no claim of completeness of coverage of the topic. I encourage you to read the referenced resources to learn more.

In the next and final chapter, we will explore how the current LLM technology fits into humanity's pursuit of artificial intelligence and what are the implications of LLMs for the human society's future.

6.10 Further learning resources

- Some foundational LLM papers:

    - *Attention is all you need* [Vas+17];
    - GPT-1 [Rad+18];
    - GPT-2 [Rad+19];
    - GPT-3 [Bro+20];
    - GPT-4 [Ach+23];
    - GPT-4o [Hur+24];
    - GPT-4o mini [Ope24];
    - O3 and O4-mini (mainly evaluation-focused) [Ope25b];
    - Llama [Tou+23b];
    - Llama 2 [Tou+23a];
    - Llama 3 [Gra+24].
    - DeepSeek-V3 [Liu+24].

 – DeepSeek-R1 [Guo+25].
 – Qwen2.5 [Yan+25a].
 – Qwen3 [Yan+25b].
 – Claude 3 [Ant24b].
 – Kimi K2 [Kim25].

- Andrej Karpathy's 4-hour tutorial *Let's reproduce GPT-2 (124M)*: https://www.youtube.com/watch?v=l8pRSuU81PU.
- Multi-modal net papers:

 – Vision transformer [Dos+20].
 – Diffusion models for image / video generation [Rom+22; Bal+24].
 – Vision-language-action models [Bro+23].
 – Audio-text models [Rad+23; Déf+24].

- Reinforcement learning [SB+98; KWW22].

 – RL applications to LLMs [Ouy+22; Sha+24].
 – RL applications outside of LLMs [Abr+24; Sil+16; Mni+15].

- Hugging Face BPE tutorial: https://huggingface.co/learn/llm-course/en/chapter6/5.

See the *Towards a commercial-grade LLM* section from earlier in the chapter for more targeted pointers and references.

6.11 References

[Abr+24]   Josh Abramson et al. "Accurate structure prediction of biomolecular interactions with AlphaFold 3". In: *Nature* 630.8016 (2024), pp. 493–500.

[Ach+23]   Josh Achiam et al. "GPT-4 technical report". In: *arXiv preprint arXiv:2303.08774* (2023).

[Ain+23]   Joshua Ainslie et al. "GQA: Training generalized multi-query transformer models from multi-head checkpoints". In: *arXiv preprint arXiv:2305.13245* (2023).

[Ala+22]   Jean-Baptiste Alayrac et al. "Flamingo: A visual language model for few-shot learning". In: *NeurIPS* 35 (2022), pp. 23716–23736.

[An+24a]   Chenxin An et al. "Training-free long-context scaling of large language models". In: *arXiv preprint arXiv:2402.17463* (2024).

[An+24b]   Haozhe An et al. "Do Large Language Models Discriminate in Hiring Decisions on the Basis of Race, Ethnicity, and Gender?" In: *arXiv preprint arXiv:2406.10486* (2024).

[Ant24a]   Anthropic. *Introducing the Model Context Protocol.* 2024. URL: https://www.anthropic.com/news/model-context-protocol (visited on 08/17/2025).

[Ant24b]   Anthropic. *The Claude 3 Model Family: Opus, Sonnet, Haiku.* https://www-cdn.anthropic.com/de8ba9b01c9ab7cbabf5c33b8 0b7bbc618857627/Model_Card_Claude_3.pdf. Model card, accessed 2025-07-13. 2024.

[Ant25]   Anthropic. *System Card: Claude Opus 4 & Claude Sonnet 4.* Tech. rep. Published May 25, 2025; accessed July 24, 2025. Anthropic, 2025. URL: https://www-cdn.anthropic.com/4263b 940cabb546aa0e3283f35b686f4f3b2ff47.pdf.

[BKH16]   Jimmy Lei Ba, Jamie Ryan Kiros, and Geoffrey E Hinton. "Layer normalization". In: *arXiv preprint arXiv:1607.06450* (2016).

[Bai+22a]   Yuntao Bai et al. "Constitutional AI: Harmlessness from AI feedback". In: *arXiv preprint arXiv:2212.08073* (2022).

[Bai+22b]   Yuntao Bai et al. "Training a helpful and harmless assistant with reinforcement learning from human feedback". In: *arXiv preprint arXiv:2204.05862* (2022).

[Bal+24]   Jason Baldridge et al. "Imagen 3". In: *arXiv preprint arXiv:2408.07009* (2024).

[Ben+03]   Yoshua Bengio et al. "A neural probabilistic language model". In: *Journal of Machine Learning Research* 3.Feb (2003), pp. 1137–1155.

[Bet+23]   James Betker et al. *Improving image generation with better captions.* 2023. URL: https://cdn.openai.com/papers/dall-e-3 .pdf.

[Bet+25]   Jan Betley et al. "Emergent misalignment: Narrow finetuning can produce broadly misaligned LLMs". In: *arXiv preprint arXiv:2502.17424* (2025).

[Blo25]   Bloomberg. *Zuckerberg is Personally Recruiting New "Super-intelligence" AI Team at Meta*. Accessed 2025-07-23. 2025. URL: https://www.bloomberg.com/news/articles/2025-06-10/zuckerberg-recruits-new-superintelligence-ai-group-at-meta.

[Bro+23]   Anthony Brohan et al. "RT-2: Vision-language-action models transfer web knowledge to robotic control". In: *arXiv preprint arXiv:2307.15818* (2023).

[Bro+20]   Tom Brown et al. "Language models are few-shot learners". In: *arXiv preprint arXiv:2005.14165* (2020). URL: https://arxiv.org/pdf/2005.14165.

[Bur25]   Theo Burman. "DeepSeek AI Refuses To Criticize Xi Jinping: "Talk About Something Else"". In: *Newsweek* (2025). Accessed 2025-07-07. URL: https://www.newsweek.com/deepseek-ai-china-tiananmen-square-xi-jinping-2021298.

[Cala]   USDC N.D. Cal. *Bartz et al. v. Anthropic PBC*. No. 3:24-cv-05417 (N.D. Cal. June 23, 2025). Order granting summary judgment on fair use; William Alsup, J. URL: https://www.courtlistener.com/docket/69058235/bartz-v-anthropic-pbc/.

[Calb]   USDC N.D. Cal. *Kadrey et al. v. Meta Platforms, Inc.* No. 3:23-cv-03417-VC (N.D. Cal. June 25, 2025). Order granting Meta's cross-motion for partial summary judgment on fair use; Vince Chhabria, J. URL: https://law.justia.com/cases/federal/district-courts/california/candce/3%3A2023cv03417/415175/598/.

[Cha25]   Avi Chawla. *Function calling & MCP for LLMs*. Daily Dose of Data Science Blog. 2025. URL: https://www.dailydoseofds.com/p/function-calling-mcp-for-llms/.

[Che+25]   Aili Chen et al. "MiniMax-M1: Scaling Test-Time Compute Efficiently with Lightning Attention". In: *arXiv preprint arXiv:2506.13585* (2025).

[Chi+25]   Wayne Chi et al. "Copilot Arena: A platform for code LLM evaluation in the wild". In: *arXiv preprint arXiv:2502.09328* (2025).

[Chi+19]   Rewon Child et al. "Generating long sequences with sparse transformers". In: *arXiv preprint arXiv:1904.10509* (2019).

[Cla+22]   Jonathan H Clark et al. "Canine: Pre-training an efficient tokenization-free encoder for language representation". In: *Transactions of the Association for Computational Linguistics* 10 (2022), pp. 73–91.

[Clo+25]   Alex Cloud et al. "Subliminal Learning: Language models transmit behavioral traits via hidden signals in data". In: *arXiv preprint arXiv:2507.14805* (2025).

[Dao+22]   Tri Dao et al. "FlashAttention: Fast and memory-efficient exact attention with IO-awareness". In: *NeurIPS* 35 (2022), pp. 16344–16359.

[DKJ25]   Bhishma Dedhia, Yuval Kansal, and Niraj K. Jha. *Bottom-up Domain-specific Superintelligence: A Reliable Knowledge Graph is What We Need.* 2025. arXiv: 2507.13966. URL: https://arxiv.org/abs/2507.13966.

[Déf+24]   Alexandre Défossez et al. "Moshi: A speech-text foundation model for real-time dialogue". In: *arXiv preprint arXiv:2410.00037* (2024).

[Det+23]   Tim Dettmers et al. "QLoRA: Efficient finetuning of quantized LLMs". In: *NeurIPS* 36 (2023), pp. 10088–10115.

[Dev+19]   Jacob Devlin et al. "BERT: Pre-training of deep bidirectional transformers for language understanding". In: *ACL NAACL-HLT.* 2019, pp. 4171–4186.

[Dos+20]   Alexey Dosovitskiy et al. "An image is worth 16x16 words: Transformers for image recognition at scale". In: *arXiv preprint arXiv:2010.11929* (2020).

[Fen+24]   Leo Feng et al. "Were RNNs all we needed?" In: *arXiv preprint arXiv:2410.01201* (2024).

[Fra+22]   Elias Frantar et al. "GPTQ: Accurate post-training quantization for generative pre-trained transformers". In: *arXiv preprint arXiv:2210.17323* (2022).

[Fuj+25]   Kazuki Fujii et al. "Rewriting pre-training data boosts LLM performance in math and code". In: *arXiv preprint arXiv:2505.02881* (2025).

[Glo+24]   Fabian Gloeckle et al. "Better & faster large language models via multi-token prediction". In: *arXiv preprint arXiv:2404.19737* (2024).

[Gra24]   Nico Grant. "Google Chatbot's A.I. Images Put People of Color in Nazi-Era Uniforms". In: *The New York Times* (2024). Accessed 2025-07-07. URL: https://www.nytimes.com/2024/0 2/22/technology/google-gemini-german-uniforms.html.

[Gra+24]   Aaron Grattafiori et al. "The Llama 3 herd of models". In: *arXiv preprint arXiv:2407.21783* (2024).

[GD23]   Albert Gu and Tri Dao. "Mamba: Linear-time sequence modeling with selective state spaces". In: *arXiv preprint arXiv:2312.00752* (2023).

[Gu+23]   Yuxian Gu et al. "MiniLLM: Knowledge distillation of large language models". In: *arXiv preprint arXiv:2306.08543* (2023).

[Guo+25]   Daya Guo et al. "Deepseek-R1: Incentivizing reasoning capability in LLMs via reinforcement learning". In: *arXiv preprint arXiv:2501.12948* (2025).

[He+16]   Kaiming He et al. "Deep residual learning for image recognition". In: *IEEE CVPR*. 2016, pp. 770–778.

[HG16]   Dan Hendrycks and Kevin Gimpel. "Gaussian error linear units (GELUs)". In: *arXiv preprint arXiv:1606.08415* (2016).

[Hof+22]   Jordan Hoffmann et al. "Training compute-optimal large language models". In: *arXiv preprint arXiv:2203.15556* (2022).

[Hol+19]   Ari Holtzman et al. "The curious case of neural text degeneration". In: *arXiv preprint arXiv:1904.09751* (2019).

[HTH25]   Kelly Hong, Anton Troynikov, and Jeff Huber. *Context Rot: How Increasing Input Tokens Impacts LLM Performance*. Tech. rep. Chroma, 2025. URL: https://research.trychroma.com/con text-rot.

[Hou+19]   Neil Houlsby et al. "Parameter-efficient transfer learning for NLP". In: *ICML*. PMLR. 2019, pp. 2790–2799.

[HR18]   Jeremy Howard and Sebastian Ruder. "Universal language model fine-tuning for text classification". In: *arXiv preprint arXiv:1801.06146* (2018).

[Hsi+24]   Cheng-Ping Hsieh et al. "RULER: What's the Real Context Size of Your Long-Context Language Models?" In: *arXiv preprint arXiv:2404.06654* (2024).

[Hu+22]   Edward J Hu et al. "LoRA: Low-rank adaptation of large language models". In: *ICLR* 1.2 (2022), p. 3.

[Hum23]   Michael Humor. *What are the LLaMA model weights?* 2023. URL: https://blog.gopenai.com/what-are-the-llama-model-wei ghts-e83a58cef1be (visited on 07/13/2025).

[Hur+24]   Aaron Hurst et al. "GPT-4o system card". In: *arXiv preprint arXiv:2410.21276* (2024).

[Hus88]   Hustler Magazine, Inc. v. Falwell. *485 U.S. 46.* 1988. URL: https://supreme.justia.com/cases/federal/us/485/46/.

[Hwa+24]   Jyh-Jing Hwang et al. "EMMA: End-to-end multimodal model for autonomous driving". In: *arXiv preprint arXiv:2410.23262* (2024).

[ICA18]   Geoffrey Irving, Paul Christiano, and Dario Amodei. "AI safety via debate". In: *arXiv preprint arXiv:1805.00899* (2018).

[Jia+24]   Albert Q Jiang et al. "Mixtral of experts". In: *arXiv preprint arXiv:2401.04088* (2024).

[Jia+23]   Albert Q. Jiang et al. *Mistral 7B.* 2023. arXiv: 2310.06825 [cs.CL].

[Kap+20]   Jared Kaplan et al. "Scaling laws for neural language models". In: *arXiv preprint arXiv:2001.08361* (2020).

[Kim25]   Kimi Team. *Kimi K2: Open Agentic Intelligence.* 2025. URL: https://github.com/MoonshotAI/Kimi-K2 (visited on 07/23/2025).

[KWW22]   Mykel J Kochenderfer, Tim A Wheeler, and Kyle H Wray. *Algorithms for Decision Making.* MIT Press, 2022.

[Kor+25]   Tomek Korbak et al. "Chain of Thought Monitorability: A New and Fragile Opportunity for AI Safety". In: *arXiv preprint arXiv:2507.11473* (2025).

[Kor+23]   Vijay Anand Korthikanti et al. "Reducing activation recomputation in large transformer models". In: *MLSys* 5 (2023), pp. 341–353.

[Kud18]   Taku Kudo. "Subword regularization: Improving neural net-
            work translation models with multiple subword candidates".
            In: *arXiv preprint arXiv:1804.10959* (2018).

[Lab+25]   Inception Labs et al. "Mercury: Ultra-Fast Language Mod-
            els Based on Diffusion". In: *arXiv preprint arXiv:2506.17298*
            (2025).

[Lam25]   Nathan Lambert. *DeepSeek V3 and the actual cost of training
            frontier AI models.* Accessed 2025-07-13. 2025. URL: https://w
            ww.interconnects.ai/p/deepseek-v3-and-the-actual-cost-of.

[LKM23]   Yaniv Leviathan, Matan Kalman, and Yossi Matias. "Fast in-
            ference from transformers via speculative decoding". In: *ICML.*
            PMLR. 2023, pp. 19274–19286.

[Lew+20]   Patrick Lewis et al. "Retrieval-augmented generation for
            knowledge-intensive NLP tasks". In: *NeurIPS* 33 (2020),
            pp. 9459–9474.

[Li+24]   Bo Li et al. "LLaVA-OneVision: Easy visual task transfer". In:
            arXiv preprint arXiv:2408.03326 (2024).

[Li+18]   Hao Li et al. "Visualizing the loss landscape of neural nets".
            In: *NeurIPS* 31 (2018).

[Lie25]   Liedtke, Michael and The Associated Press. *AI kingpin Nvidia
            crowned as first public company with a $4 trillion valuation.*
            July 9, 2025. URL: https://apnews.com/article/nvidia-4-trillio
            n-chipmaker-7947e86a7ee9a994b9f16c3c0779b74f (visited on
            07/13/2025).

[Lin+24]   Ji Lin et al. "AWQ: Activation-aware weight quantization for
            on-device LLM compression and acceleration". In: *MLSys* 6
            (2024), pp. 87–100.

[Liu+24]   Aixin Liu et al. "DeepSeek-V3 technical report". In: *arXiv
            preprint arXiv:2412.19437* (2024).

[MXC24]   Yu Meng, Mengzhou Xia, and Danqi Chen. "SimPO: Sim-
            ple preference optimization with a reference-free reward". In:
            NeurIPS 37 (2024), pp. 124198–124235.

[Met25]   Meta. *The Llama 4 herd: The beginning of a new era of natively
            multimodal AI innovation.* 2025. URL: https://ai.meta.com/bl
            og/llama-4-multimodal-intelligence/ (visited on 07/23/2025).

[Mik13]   Tomas Mikolov. "Efficient estimation of word representations in vector space". In: *arXiv preprint arXiv:1301.3781* 3781 (2013).

[Mni+15]   Volodymyr Mnih et al. "Human-level control through deep reinforcement learning". In: *Nature* 518.7540 (2015), pp. 529–533.

[Nas+25]   Ali Naseh et al. "R1dacted: Investigating Local Censorship in DeepSeek's R1 Language Model". In: *arXiv preprint arXiv:2505.12625* (2025).

[Nie+25]   Shen Nie et al. "Large language diffusion models". In: *arXiv preprint arXiv:2502.09992* (2025).

[Ope24]   OpenAI. *GPT-4o mini: Advancing cost-efficient intelligence.* 2024. URL: https://openai.com/index/gpt-4o-mini-advancing-cost-efficient-intelligence/.

[Ope25a]   OpenAI. *Introducing gpt-oss: gpt-oss-120b and gpt-oss-20b push the frontier of open-weight reasoning models.* 2025. URL: https://openai.com/index/introducing-gpt-oss/ (visited on 08/05/2025).

[Ope25b]   OpenAI. *OpenAI o3 and o4-mini System Card.* Tech. rep. OpenAI, 2025.

[Ope25c]   OpenAI. *What are tokens and how to count them?* https://help.openai.com/en/articles/4936856-what-are-tokens-and-how-to-count-them. Updated 1/28/2025; accessed 2025-07-13. 2025.

[Ouy+22]   Long Ouyang et al. "Training language models to follow instructions with human feedback". In: *NeurIPS* 35 (2022), pp. 27730–27744.

[Par+23]   Joon Sung Park et al. "Generative agents: Interactive simulacra of human behavior". In: *ACM UIST.* 2023, pp. 1–22.

[Pen+23]   Bowen Peng et al. "Yarn: Efficient context window extension of large language models". In: *arXiv preprint arXiv:2309.00071* (2023).

[PSM14]   Jeffrey Pennington, Richard Socher, and Christopher D Manning. "GloVe: Global vectors for word representation". In: *EMNLP.* 2014, pp. 1532–1543.

[Pol+23]   Michael Poli et al. "Hyena hierarchy: Towards larger convolutional language models". In: *ICML*. PMLR. 2023, pp. 28043–28078.

[PW16]   Ofir Press and Lior Wolf. "Using the output embedding to improve language models". In: *arXiv preprint arXiv:1608.05859* (2016).

[Pre+22]   Ofir Press et al. "Measuring and narrowing the compositionality gap in language models". In: *arXiv preprint arXiv:2210.03350* (2022).

[RS21]   Markus N Rabe and Charles Staats. "Self-attention does not need $O(n^2)$ memory". In: *arXiv preprint arXiv:2112.05682* (2021).

[Rad+18]   Alec Radford et al. "Improving language understanding by generative pre-training". In: (2018).

[Rad+19]   Alec Radford et al. "Language models are unsupervised multitask learners". In: *OpenAI blog* 1.8 (2019), p. 9.

[Rad+23]   Alec Radford et al. "Robust speech recognition via large-scale weak supervision". In: *ICML*. PMLR. 2023, pp. 28492–28518.

[Raf+23]   Rafael Rafailov et al. "Direct preference optimization: Your language model is secretly a reward model". In: *NeurIPS* 36 (2023), pp. 53728–53741.

[Raf+19]   Colin Raffel et al. "Exploring the Limits of Transfer Learning with a Unified Text-to-Text Transformer". In: *arXiv e-prints* (2019). arXiv: 1910.10683.

[RMY24]   Abdelrahman Ragab, Mohammad Mannan, and Amr Youssef. ""Trust Me Over My Privacy Policy": Privacy Discrepancies in Romantic AI Chatbot Apps". In: *IEEE EuroS&PW*. IEEE. 2024, pp. 484–495.

[Raj+20]   Samyam Rajbhandari et al. "ZeRO: Memory optimizations toward training trillion parameter models". In: *SC20: International Conference for High Performance Computing, Networking, Storage and Analysis*. IEEE. 2020, pp. 1–16.

[Raj+25]   Meghana Rajeev et al. *Cats Confuse Reasoning LLM: Query Agnostic Adversarial Triggers for Reasoning Models*. 2025. arXiv: 2503.01781. URL: https://arxiv.org/abs/2503.01781.

[Rob+24]   David Robinson et al. "NatureLM-audio: An audio-language foundation model for bioacoustics". In: *arXiv preprint arXiv:2411.07186* (2024).

[Rom+22]   Robin Rombach et al. "High-resolution image synthesis with latent diffusion models". In: *IEEE/CVF CVPR*. 2022, pp. 10684–10695.

[Sah+24]   Pranab Sahoo et al. "A systematic survey of prompt engineering in large language models: Techniques and applications". In: *arXiv preprint arXiv:2402.07927* (2024).

[San+19]   Victor Sanh et al. "DistilBERT, a distilled version of BERT: Smaller, faster, cheaper and lighter". In: *arXiv preprint arXiv:1910.01108* (2019).

[SH96]   Jürgen Schmidhuber and Stefan Heil. "Sequential neural text compression". In: *IEEE Transactions on Neural Networks* 7.1 (1996), pp. 142–146.

[SHB15]   Rico Sennrich, Barry Haddow, and Alexandra Birch. "Neural machine translation of rare words with subword units". In: *arXiv preprint arXiv:1508.07909* (2015).

[Sha+24]   Zhihong Shao et al. "DeepSeekMath: Pushing the limits of mathematical reasoning in open language models". In: *arXiv preprint arXiv:2402.03300* (2024).

[Shi24]   Anton Shilov. *Datacenter GPU Service Life Can Be Surprisingly Short – Only One to Three Years Is Expected According to Unnamed Google Architect*. 2024. URL: https://www.tom shardware.com/pc-components/gpus/datacenter-gpu-servic e-life-can-be-surprisingly-short-only-one-to-three-years-is-e xpected-according-to-unnamed-google-architect (visited on 07/13/2025).

[Shi+23]   Noah Shinn et al. "Reflexion: Language agents with verbal reinforcement learning". In: *NeurIPS* 36 (2023), pp. 8634–8652.

[Sho+19]   Mohammad Shoeybi et al. "Megatron-LM: Training multi-billion parameter language models using model parallelism". In: *arXiv preprint arXiv:1909.08053* (2019).

[Sil+16]   David Silver et al. "Mastering the game of Go with deep neural networks and tree search". In: *Nature* 529.7587 (2016), pp. 484–489.

[Sta+24]   Elizabeth C Stade et al. "Large language models could change the future of behavioral healthcare: A proposal for responsible development and evaluation". In: *NPJ Mental Health Research* 3.1 (2024), p. 12.

[Su+24]   Jianlin Su et al. "RoFormer: Enhanced transformer with rotary position embedding". In: *Neurocomputing* 568 (2024), p. 127063.

[SKU24]   Jingtong Su, Julia Kempe, and Karen Ullrich. "Mission impossible: A statistical perspective on jailbreaking llms". In: *Advances in Neural Information Processing Systems* 37 (2024), pp. 38267–38306.

[SB+98]   Richard S Sutton, Andrew G Barto, et al. *Reinforcement Learning: An Introduction*. Vol. 1. 1. MIT Press, 1998.

[Tay+21]   Yi Tay et al. "Charformer: Fast character transformers via gradient-based subword tokenization". In: *arXiv preprint arXiv:2106.12672* (2021).

[Ton+25]   Chengzhuo Tong et al. "Delving into RL for Image Generation with CoT: A Study on DPO vs. GRPO". In: *arXiv preprint arXiv:2505.17017* (2025).

[Tou+23a]   Hugo Touvron et al. "Llama 2: Open foundation and fine-tuned chat models". In: *arXiv preprint arXiv:2307.09288* (2023).

[Tou+23b]   Hugo Touvron et al. "Llama: Open and efficient foundation language models". In: *arXiv preprint arXiv:2302.13971* (2023).

[Tye24]   Jordyn C Tye. "Exploring the Intersections of Privacy and Generative AL: A Dive into Attorney-Client Privilege and ChatGPT". In: *Jurimetrics* (2024), pp. 309–40.

[Vas+17]   Ashish Vaswani et al. "Attention is All You Need". In: *NeurIPS*. 2017. URL: https://arxiv.org/pdf/1706.03762.pdf.

[Wan+25]   Guan Wang et al. *Hierarchical Reasoning Model*. 2025. arXiv: 2506.21734. URL: https://arxiv.org/abs/2506.21734.

[Wan+24a]   Pengkun Wang et al. "LLM-AutoDA: Large language model-driven automatic data augmentation for long-tailed problems". In: *NeurIPS* 37 (2024), pp. 64915–64941.

[Wan+23]   Weizhi Wang et al. "Augmenting language models with long-term memory". In: *NeurIPS* 36 (2023), pp. 74530–74543.

[Wan+24b]   Xindi Wang et al. "Beyond the limits: A survey of techniques to extend the context length in large language models". In: *arXiv preprint arXiv:2402.02244* (2024).

[Wan+24c]   Xingyao Wang et al. "Executable code actions elicit better llm agents". In: *ICML*. 2024.

[Wei+22a]   Jason Wei et al. "Chain-of-thought prompting elicits reasoning in large language models". In: *NeurIPS* 35 (2022), pp. 24824–24837.

[Wei+22b]   Jason Wei et al. "Emergent abilities of large language models". In: *arXiv preprint arXiv:2206.07682* (2022).

[Wika]   Wikipedia. *AutoGPT*. URL: https://en.wikipedia.org/wiki/AutoGPT (visited on 07/14/2025).

[Wikb]   Wikipedia. *Bradley-Terry model*. URL: https://en.wikipedia.org/wiki/Bradley%E2%80%93Terry_model (visited on 05/30/2025).

[Wikc]   Wikipedia. *JSON*. URL: https://en.wikipedia.org/wiki/JSON (visited on 08/17/2025).

[Wikd]   Wikipedia. *LangChain*. URL: https://en.wikipedia.org/wiki/LangChain (visited on 07/14/2025).

[Wike]   Wikipedia. *Self-supervised learning*. URL: https://en.wikipedia.org/wiki/Self-supervised_learning (visited on 07/29/2025).

[Wor+22]   Mitchell Wortsman et al. "Model soups: Averaging weights of multiple fine-tuned models improves accuracy without increasing inference time". In: *ICML*. PMLR. 2022, pp. 23965–23998.

[Xie+25]   Yichen Xie et al. "S4-Driver: Scalable Self-Supervised Driving Multimodal Large Language Model with Spatio-Temporal Visual Representation". In: *IEEE/CVF CVPR*. 2025, pp. 1622–1632.

[Xio+23]   Wenhan Xiong et al. "Effective long-context scaling of founda-
           tion models". In: *arXiv preprint arXiv:2309.16039* (2023).

[Xu+24]   Xiaohan Xu et al. "A survey on knowledge distillation of large
           language models". In: *arXiv preprint arXiv:2402.13116* (2024).

[Xue+22]   Linting Xue et al. "ByT5: Towards a token-free future with
           pre-trained byte-to-byte models". In: *Transactions of the As-
           sociation for Computational Linguistics* 10 (2022), pp. 291–
           306.

[Yan+25a] An Yang et al. "Qwen2.5 Technical Report". In: *arXiv preprint
           arXiv:2412.15115* (2025). URL: https://arxiv.org/abs/2412.15
           115.

[Yan+25b] An Yang et al. "Qwen3 technical report". In: *arXiv preprint
           arXiv:2505.09388* (2025).

[Yao+23]   Shunyu Yao et al. "Tree of thoughts: Deliberate problem
           solving with large language models". In: *NeurIPS* 36 (2023),
           pp. 11809–11822.

[Yu+24]   Zhiyuan Yu et al. "Don't listen to me: Understanding and
           exploring jailbreak prompts of large language models". In:
           USENIX Security 24. 2024, pp. 4675–4692.

[Yua+25]   Ruibin Yuan et al. "Yue: Scaling open foundation mod-
           els for long-form music generation". In: *arXiv preprint
           arXiv:2503.08638* (2025).

[Zha+19]   Jingzhao Zhang et al. "Why gradient clipping accelerates
           training: A theoretical justification for adaptivity". In: *arXiv
           preprint arXiv:1905.11881* (2019).

[Zho+22]   Yanqi Zhou et al. "Mixture-of-experts with expert choice rout-
           ing". In: *NeurIPS* 35 (2022), pp. 7103–7114.

[Zho+24]   Zhenhong Zhou et al. "How alignment and jailbreak work:
           Explain LLM safety through intermediate hidden states". In:
           arXiv preprint arXiv:2406.05644 (2024).

7 In pursuit of artificial intelligence

In this final chapter, we address how large language models (LLMs) fit into humanity's pursuit of the human-grade mind in the machine, also known as artificial intelligence (AI). We will touch on a variety of issues, including:

- What is AI?
- How do LLMs compare to human intelligence?
- What are the opportunities and risks AI carries for the human society?

The discussion is grounded in research, however, due to the fast-evolving nature of this field, it does involve an unavoidable degree of speculation.

7.1 What is AI?

Throughout this book you have *experienced* what artificial intelligence (AI) actually is from the perspective of contemporary computer science. Now it is time to try to define it. There are many definitions of AI. Apparently, the term was originally introduced in 1955 by John McCarthy in a research proposal [McC+06], with the following definition:

> "...the artificial intelligence problem is taken to be that of making a machine behave in ways that would be called intelligent if a human were so behaving."

> (Archive copy of the text is available here: http://jmc.stanford .edu/articles/dartmouth/dartmouth.pdf.)

However, this definition of artificial intelligence is self-referential – it itself relies on the definition of human intelligence. Over time, I have come up with the following more direct definition that I prefer:

> *Artificial intelligence (AI) is a human-engineered system capable of making decisions based on information.*

This definition of artificial intelligence covers, for example, tasks like playing chess, recommending movies, predicting weather, translating languages, or, indeed, generating new text LLM-style. In fact, one could argue that the earliest mechanical lock able to decide the correct key shape from the incorrect one was an early form of artificial intelligence; it implemented a physical decision function, albeit in metal rather than in silicon.

However, colloquially, we would not call a lock AI. At this point, even services like Google Translate, Google search, or TikTok recommender engine might not be called AI. This illustrates the phenomenon of moving goalposts – once a task becomes solvable with software, that activity no longer feels like "true intelligence" or "real thinking." Some call this the "AI effect" [HK19; MC04] – we dismiss machine intelligence when it becomes effective.

Other terms that frequently pop up in AI discussions include *general artificial intelligence* and *superintelligence*. While definitions of these terms are also not universally agreed upon, I understand them as follows: In contrast to *narrow* AI, which is specialized for a specific task, *general* AI is a system that would be able to handle any intellectual task a human can; and *superintelligence* is intelligence that exceeds humans in many / all areas.

7.2 How do LLMs and humans compare?

Specialized AI surpassing humans on narrow outcome-per-time tasks is not new. AI now defeats human champions in the games of Go [Sil+16] and chess [Sil+18]; processes huge volumes of data to return search or recommendation results [Liu+22; CAS16; Sun+19; Gen+22]; and can make superhuman predictions about protein-folding structure [Abr+24]; among many other feats.

Large language models (LLMs), however, are the first class of AI that approaches – or in some domains exceeds – human performance across a genuinely general range of cognitive tasks rendered in text. To better understand the landscape, we can compare LLMs and human intelligence along multiple dimensions.

Short-term memory. While humans have an estimated short-term memory capacity of 7 ± 2 items [Mil56], many modern LLMs can simultaneously attend to context windows of 128 thousand tokens [Ope24; Gra+24]. This allows an LLM to consume huge volumes of text at a speed much faster than a human could ever manage.

Utilization rate. AI models can operate continuously, 24/7. Humans, by contrast, typically work only about 40 hours out of the full week ($7 \cdot 24 = 168$ hours); that leaves 168 - 40 = 128 hours – roughly $128/168 \approx 76\%$ of the week – spent outside of work.

Training duration. It takes around 18 years to produce a human worker, and potentially as long as an extra decade to make an expert in the field (e.g., Bachelor's + PhD degrees). In turn, AI training duration ranges in time from weeks to months – for example, 21 days reported for Llama 1 [Tou+23] and (estimated) several months for Llama 3 [Sha24; Gra+24]. DeepSeek-V3 also took around 2 months to train [Liu+24].

Copying ease. Creating an LLM copy is instant – just copy over the weights of the model, at effectively zero marginal cost. Human brains sadly have not evolved such a direct copying mechanism, and so human duplication requires growing one from scratch, which is much less scalable and much more costly. By the way, this ease of duplication is why an AI will probably out-survive humanity – LLM files could be beamed at the speed of light into space to a suitable receiver, escaping whatever disaster becomes the planet Earth. That said, future improvements in understanding of human brain neuroplasticity could perhaps one day allow for something like copying, akin to an observed transfer of brain function between brain areas under external stimulus [Ser+20].

Speed of improvement. Measured from the 2017 introduction of the transformer architecture [Vas+17], LLMs' rise has been astonishingly rapid relative to the usual cadence of technological progress in human history. In this time, LLMs have gone from outputting error-filled gibberish to being able to keep up a PhD-expert-level conversation. Humans' biology, in turn, takes orders of magnitude more time to evolve naturally. While biological engineering and DNA editing technologies could theoretically enhance the rate of human biological "wetware" improvement, we are currently nowhere near to closing the gap with the speed of AI advances.

Energy consumption. While a lot of attention has been focused on how much energy is consumed by AI training [De 23], it turns out that things are not so clear cut when we compare an LLM vs. a human brain at inference on a standardized basis. For example, when compared on the *energy expenditure per question* basis, a model like GPT-4o mini seems to be more energy-efficient than a human for some question types [Bil24]. Even on the aggregate, for now AI represents a small fraction of total data center energy consumption, and there are reasons to think that the dramatic forecasts about AI's energy consumption are an overestimate [Cas24].

Number of parameters. We have no true understanding of how neuron firing in a human brain brings about intelligence. Yet, we can compare some numbers for fun. A human brain contains around 86 billion neurons [Aze+09] and on the order of hundreds of trillions of synapse connection between neurons [Dra05]. As to LLMs, for instance, DeepSeek-V3 model has 671 billion total parameters, with 37 billion parameters activated on each token due to the use of mixture-of-experts architecture [Liu+24]. Kimi K2 model has 1.04 trillion parameters with 32.6B activated per token [Kim25]. GPT-4 has been rumored to have around 1.76 trillion parameters [Sch23]. So LLMs are increasing in numerical complexity to brain-like levels, however, this does not mean much as structures and algorithms taking advantage of such numerical capacity likely differ.

Mechanism. Large language models (LLMs) are trained using a simple principle – predict the next token. Remarkably, this basic task leads to models that can generate text displaying elements of human-like intelligence. This raises the intriguing possibility that prediction-based learning may also underlie human intelligence. Indeed, some researchers suggest the brain functions as a prediction machine, constantly trying to minimize surprise from sensory input [Fri10; Cla13].

Unlike LLMs though, the human brain is multi-modal first – integrating vision, sound, touch, smell, imagination, and more – and learns through embodied, real-time interaction with the world. This learning is continuous, remembered, and involves interaction with others.

While some modern LLMs are beginning to incorporate multi-modal inputs and even generate images, audio, or video, full multi-modal prediction and surprise minimization remain active research frontiers. Research integrating video prediction with text cues is active in such fields as autonomous driving, robotics, and video generation [Xie+25; Hwa+24; Bro+23; Bar+24]. Experiments on integrating long-term memory into LLMs are also underway [Wan+23]. Multi-agent LLM systems are actively being researched

[Guo+24]. It is possible that LLMs will more closely approximate human cognitive abilities once they move beyond the text-first domain and adopt richer learning modalities grounded in long-term memory and multi-agent interactions.

At the same time, there is research showing that LLMs can maintain high prediction quality while not having a robust *world model* – unlike humans [Vaf+25]. This raises a possibility there might be some fundamental difference in how reasoning is organized in LLMs vs. humans – that we don't quite understand yet – allowing for the greater relative robustness in the human world model.

Data efficiency. While human children can acquire language from input of less than 100 million words, current state-of-the-art LLMs require multiple orders of magnitude more data – on the order of trillions of words [War+23; Gil+17]. While LLM training is still much faster than human learning, this data efficiency discrepancy may hint at the limitations of the current LLM architecture and / or learning algorithms. Resolving this puzzle could enable substantial LLM miniaturization with minimal loss in performance – which is an active research area [CV25].

Ability. Since the emergence of LLMs, there has been a desire to put a number on how well humans vs. LLMs perform on a variety of tasks or outcomes, with a variety of benchmarks proposed and new benchmarks appearing every day. So far, the results of such evaluations have shown that LLMs can exceed the average human's ability and approach the expert level of ability on many cognitive tasks (with some caveats).

For example, one benchmark that has been very popular historically is *MMLU* (Massive Multitask Language Understanding) [Hen+20], which consists of thousands of multiple choice questions, 4 options each, across 57 subjects, such as science, law, religion, etc. In the evaluation, an LLM is provided with the text of the problem and the answer options and LLM's task is to predict the index for the correct option. Human experts achieved around 89.8% accuracy on this benchmark, non-experts (Amazon Mechanical Turk workers) were at around 34.5%, and GPT-3 175B – the top performing LLM in the study – achieved 43.9% accuracy at the time of MMLU release [Hen+20]. By 2024, however, most top LLM models (Llama 3, GPT-4o, Claude 3.5 Sonnet) achieved 87-89% levels on this benchmark, matching human experts; see Table 2 in [Gra+24] for the top LLM comparisons on MMLU and many, many other benchmarks as of July 2024.

On all kinds of standardized tests (LSAT, GRE, etc.), top LLMs tend to achieve 80-90% accuracy and 80th-90th percentile of human performance (or even higher) – see Table 17 in [Gra+24] and Table 1 in [Ach+23].

One fun study [McD+25] has found that among experimental conditions (1) medical doctors alone, (2) medical doctors in combination with an LLM, and (3) LLM working alone without human intervention, the latter achieved the best scores on the comprehensive differential diagnosis task (presence of a human medical doctor seems to have hurt the LLM's performance, which could be interpreted as welcome news by those hoping for LLMs not just to *complement*, but to *replace* humans).

In 2025, Google's advanced reasoning version of Gemini LLM and an experimental OpenAI reasoning LLM both earned a gold-medal score at the International Mathematical Olympiad, the world's most prestigious (and notoriously difficult) competition in mathematics for high school students; each model correctly solved 5 of 6 problems under standard contest conditions; the competition this year included 630 participants from 110 countries (maximum of 6 top high-schoolers from each participating country), of whom 67 achieved gold ($\sim 11\%$) [IMO25; LL25; Var25]. This achievement puts these LLMs squarely at the level of world's top young mathematicians. Both LLMs remained unreleased to the public as the wins were announced.

Figure 7.1: SVG image generated by OpenAI's o4-mini LLM in response to the prompt "generate an SVG of a pelican riding a bicycle" (July 2025)

The subjective assessments of the LLMs by experienced LLM users – so-called *vibes* – are also an important litmus test of the model quality. One fun example of this is Simon Willison's quick test of intelligence – asking an LLM to "generate an SVG of a pelican riding a bicycle" [Cla25; Wil25]. (SVG, or Scalable Vector Graphics, is a *vector* image format, where the image is specified by mathematical equations that describe shapes, lines, and curves, rather than by pixel grids [Wikb], so the image can be rendered at an arbitrary resolution without quality degradation.) See Figure 7.1 for the corresponding SVG solution generated by OpenAI's `o4-mini` . Being able to describe an imagined object's appearance using mathematical formulas like this is an impressive feat. Beyond pelicans, SVG image generation can be useful, for example, in floorplan design [SHF23]. An extension of this idea are LLM-based models that can generate 3D assets, for example, for animation or product design [Lu+25].

However, these evaluation results deserve to be interpreted with some caution. A sneaky issue is the problem of *training data contamination* [Roo24] – it is possible that LLMs are trained on the very questions they get evaluated on (or on some very similar versions of those questions). In case of such contamination, LLM performance may be reflective of answer memorization rather than true deep reasoning. It may not be as big of a problem if we think about an LLM replacing a human worker whose main job is essentially memory retrieval (as is arguably the case for some workers in the medical and legal fields), but it is much more problematic if we think about original research, innovation, and problem solving.

The fact that model performance has plateaued / saturated [Wan+24b] on many of the benchmarks, such as MMLU, does not alleviate the worry. As a response, more challenging benchmarks have been proposed, such as GPQA [Rei+24] and MMLU-Pro [Wan+24b], which, over time, also showed signs of improving LLM performance and saturation. This could be because LLM creators figured out where LLMs were not doing well and, for example, added chain-of-thought processing or other tricks. Or it could be that bad actors just overfit the benchmark; this is a real worry, considering there is a huge incentive to cheat – if an LLM suddenly gets ranked high, it may mean huge money flows for its creators, through new investments and / or new users. This situation is appropriately characterized by Goodhart's law [Wika], which is usually stated as "when a measure becomes a target, it ceases to be a good measure."

As a result of the possible unintentional or intentional contamination-by-benchmark in LLM training, unfortunately, all the public benchmark evaluations should be taken with a grain of salt.

Fundamentally, there are several possible ways of mitigating the contamination issue:

- Create new tough evaluation task sets and keep them a well-guarded secret [Ope25; Kut+25]. For example, on the unpublished advanced math problem set FrontierMath from [Gla+24], a top LLM achieved only 2% success rate (with Claude 3.5 Sonnet, Gemini 1.5 Pro, and GPT-4o evaluated). See https://epoch.ai/frontiermath for the current leaderboard. I would venture to say it is now an AI researcher sport to propose new benchmarks that LLMs would fail at – once no such new benchmarks can be found, perhaps we have arrived at real general AI!

- Figure out a way to auto-generate many completely new kinds of problems for evaluation [Sho+25; OL25; Var+25].

- Engage in long-form evaluation with a focus on performance on specific complex tasks [Khr+24; Zhe+23]. An example of this are the coding benchmarks, such as SWE-bench (https://www.swebench.com/) [Jim+23], which evaluates how well LLMs can resolve real-life code issues posted on GitHub; as of July 2025, the best models achieve around 33.8% resolution rate. An altered version of this benchmark (by OpenAI), called SWE-bench Verified [Cho+24], which removes some ambiguous or impossible-to-solve issues and makes other improvements, shows top models achieving around 75% resolution rate.

- Rely on extensive human-expert paired comparison evaluations across a variety of tasks – an example of this includes the crowdsourced LMArena effort (https://lmarena.ai/leaderboard) [Chi+25; Chi+24; Zhe+23].

ℹ Human labor behind AI

Outside of volunteer-based crowdsourcing, using human experts to build data sets for AI training can be expensive – but remains quintessential – and is behind the demand for human labeler services of companies like Scale AI (https://scale.com/). However, human laborers do not just help create AI training data sets; they can also play a critical role in providing verification and quality control services. As one example, if a Waymo self-driving car runs into any kind of trouble, like getting stuck without moving for some time, a remote human operator will log in into the on-board system to try to fix the situation [AAV22]. In medicine, a lot of radiology work now involves AI effectively making the diagnosis and a human doctor verifying it [Naj23]. On commercial flights, a large portion of the flight is regulated by the autopilot [Nel+98], with a human pilot being there to make sure everything looks fine; there is, in fact, a push to fully replace human pilots in the military [Cro25] and cargo flight [Pri23] applications, which may then overflow into the civil passenger aviation. This type of work of a "human nanny for AI" is likely to be in demand for some time – until AI reaches quality levels, where human supervision becomes truly pointless. Last but not least, AI developers, many of whom have PhDs, are also human laborers behind AI – countless years of PhD lives have been spent on search for the right architectures and techniques that enable the best AI performance possible and that underlie a lot of the material presented in this book, as can be seen from all the citations.

In terms of the broad conclusions that can be drawn, there are several:

- Multi-modal LLMs tend to do better than the text-only models [Wan+24a].

- Models with reasoning (multi-step text generation) tend to perform better than one-shot LLMs [Wei+22].

- There are certainly some text reasoning areas where human experts still beat LLMs [Gla+24]. Yet, there are many areas where experts underperform LLMs [Luo+25; McD+25; Sil+16]. Most critically, an average human loses to LLMs on a vast set of intellectual tests [Liu+24; Hen+20].

- On things like embodied 3D spatial reasoning and real world interactions, we could expect the multi-modal varieties of LLMs to catch up to humans at some point. Overall, there does not seem to be any insurmountable problem here other than collecting the right kind of data, where game physics simulators, for instance, could be very helpful.

A good example of the embodied AI capabilities is autonomous driving, where Waymo cars already get into accidents at *a far lower rate than human drivers on per-mile-driven basis* – so their use directly saves lives – something very hard to lobby against even if their deployment would eliminate many, many jobs [Di +24; Kus+25].

- More generally, if you worry about unpredictable failure modes of LLMs [Kic+23], it is worth remembering that humans are vulnerable to erring too. Thus, *it is not critical for LLMs to be perfect, they just need to be better than a critical mass of humans* for the demand for LLMs to grow.

Sentience. Could an LLM be sentient / conscious like humans? It is a philosophical question with no consensus, but a few points can be made:

- We do not know how to rigorously test for consciousness. Even in humans, we cannot directly access others' conscious streams; so we assume others are conscious based on our own experience and the expectation that other humans should be like us [Cha97].

- Our own judgments of sentience are heavily biased by linguistic and communication competence, which we use as a proxy signal for consciousness in other humans – and, perhaps, in animals like dogs that have evolved human-tailored facial expressions [Kam+19].

- Some researchers have proposed precautionary ethics frameworks, suggesting that if we cannot rule out an intelligent system's sentience, we may be obligated to consider their welfare and the risk of their silent suffering [Tka24].

- Even if "real" sentience in AI remains unrealized or philosophically unresolved, designing AI systems to exhibit sentient-like behavior constitutes a valid and practical optimization objective [Han+25]. And if we one day obtain AI systems that *appear* conscious based on every metric we can come up with, how could we be confident such systems are not actually conscious?

For now, AI sentience remains speculative. That said, research into AI-adjacent biological brain organoids – essentially "brains in jars" – suggests they could outperform silicon-based computers in tasks such as 3D control and robotics [Kag+22]. In fact, it is already possible to rent brain organoid compute [Fox24]. Biohybrid robots powered by living cells, such as fungi, are also just around the corner [Mis+24]. Skeptics may find it easier to entertain the possibility of sentience in such carbon-based biological units than in silicon-based digital models.

7.3 What is the future of LLMs?

Here we can speculate about some scenarios about our and LLMs' future.

7.3.1 Singularity

There is active ongoing research on AI's ability to produce research findings that improve AI itself [Liu+25; Faw+22]. One scenario that has been considered is the possibility that LLMs doing research could start self-improving to a point where we reach a singularity – a moment where the rate of self-improvement goes up exponentially and we end up with a superintelligence on our hands. This idea has been crystallized, for example, in the widely publicized report "Situational awareness" [Asc24]. The imagination draws an almost omnipotent AI that could, at a whim, make for its creator a fortune in the financial markets, find a cure for all cancer, or topple a foreign government.

The singularity idea has its roots in so-called *scaling laws* [Kap+20] – an empirically (experimentally) observed relationship in AI training, that bigger AI models trained on more data with more compute tend to perform better – and predictably so, without a clear ceiling to the pattern.

The core hypothesis underlying the singularity idea is that with infinite quality data, infinite compute, and infinitely large models, we could reach arbitrarily high quality of LLMs. In practice, though, we may have already exhausted the high-quality data – and pushing forward with more compute / larger models is very costly [She+25], so we are yet to see if the prediction will be fulfilled.

However, even if we might be skeptical, we cannot technically rule out that this scenario might be realized, given the rapid pace of AI research.

Unofficially, the belief in such singularity is alive and well among many inside AI labs – to what degree it is held sincerely vs. declared for continued fund-raising purposes is less clear.

It is also worth pointing out that the arrival of very advanced AI may not be obvious to the public for a long while. It could be so expensive to run at first that it would only be viable as a research exercise in the well-funded AI labs, not commercially viable for individual consumer use. Further, such AI could be concealed on national security and corporate secrecy grounds, while public is fed far inferior AI versions than the state of the art. (If you had any doubts about whether AI and national security domains intersect, the former US National Security Agency (NSA) chief Paul M. Nakasone joined OpenAI's board in 2024 [The24].)

7.3.2 Break-throughs in biotech

There is a possibility that the existing LLM technology is already sufficient to deliver a breakthrough in biotech given the right data.

For instance, one promising area is decoding the DNA and enabling translation of human language instructions into DNA. As a thought experiment, consider specifying a floor plan for a wooden house you want and AI encoding it for you into a DNA sequence that gets implanted into a seed that grows into a house you desire. With enough high-quality data, this might not be a much harder mathematical problem than building an LLM to translate English to Chinese.

To advance towards this vision, however, we need much better computer simulators of molecular-biological systems that could generate the required training data; works like AlphaFold, which predicts how proteins fold in 3D space based on their amino-acid sequences, are a step in this direction [Abr+24] – the Nobel Prize in Chemistry in 2024 was awarded, in part, for the corresponding research (https://www.nobelprize.org/prizes/chemistry/2024/hassabis/lecture/). Active research on LLMs for genetic sequence generation is ongoing, as illustrated by Evo 2 model [Bri+25].

My intuition is that the LLM technology applications in biotech are going to be quite dramatic.

7.3.3 Labor displacement

Businesses naturally seek to reduce labor costs if they can do so without hurting revenues, even if this means replacing humans with AI. As AI grows more capable and productive, increasing numbers of people will find themselves supplanted in the workplace.

On the one hand, such labor displacement holds a great promise. We can make expert-level medics, lawyers, educators available for cheap on tap to the whole world.

Take medicine as an example: if an AI system can diagnose and prescribe with a lower error rate than an overworked, time-pressed primary care physician, why maintain a medical monopoly on prescriptions? Primary care often depends less on genius-level insight and more on protocol adherence, rapid knowledge retrieval, and clear communication – areas in which large language models excel. Routine triage, diagnosis, treatment planning, and even mental health counseling could be handled by AI, reserving human specialists for edge cases and nuanced judgment calls.

The same principle applies to other regulated professions, such as law. For high-stakes cases, humans may insist on expert verification or a human-in-the-loop fallback. Yet as AI crosses psychological thresholds of trust and reliability, people may come to fear human involvement as much as AI error.

From this perspective, the rational path is to regulate outcomes, not credentials when it comes to licenses to practice. If an AI can meet objective standards for reliability, transparency, and liability – and match or outperform human practitioners on these criteria – it should be authorized to practice. Ending monopolies on professional credentials could dramatically expand access in underserved areas.

I am especially excited about education, where much of the university pipeline could be replaced by a video-streaming service, AI tutor, and independent assessment centers. I am waiting for YouTube to start formally offering bachelor's degrees – we are running out of excuses to continue gate-keeping free education from anyone who wants it.

Laws would probably need to be rewritten to accommodate this – for example, to ensure privileged AI conversations, to minimize ambiguity in legal texts, to regulate liability, to break up credentialing monopolies, to potentially relax intellectual property protections that might restrict AI creativity, and so on.

On the other hand, massive labor displacement can cause significant economic disruption.

While historically technological advances have often destroyed fewer jobs than they created, there is no immutable law that guarantees the same balance will hold in an AI-driven future. In fact, there is some evidence that since 1980s technological innovation has replaced more US jobs than it has generated [Aut+24].

With robotics and AI advancing rapidly, the labor of most humans – whether manual or white-collar – could become unnecessary. For instance, autonomous vehicles like Waymo could easily replace all ride-hail drivers. See [Tom+25] for a list of jobs at the highest and the lowest risk of displacement by AI.

Such a shift may force a rethinking of social organization and push policymakers toward more socialist-leaning policies. Early experiments with universal basic income (UBI) [Bid19] may be the harbingers of this change.

Paradoxically, the very goal of technological progress has always been to reduce the need for human toil – so perhaps we should embrace rather than fear this long-cherished dream.

Still, many roles will endure longer due to the difficulty of their automation – including politicians, AI developers, biologists, surgeons, and the like.

7.3.4 Personalization

The prevailing paradigm in LLM development emphasizes "alignment" with a set of pre-approved criteria – essentially ensuring that models adhere to politically correct norms – to satisfy regulators and appease certain segments of the public [Bai+22]. In contrast, many users crave the opposite – a fully personalized AI companion that indulges their whims, unconstrained by any censorial "big brother."

Thanks to the ease of fine-tuning and jailbreaking AI models, a growing number of open-weight models, and advances in model miniaturization and scaling, I anticipate a future where everyone can run their own private, completely uncensored LLM.

While this democratization of AI brings enormous benefits, it also carries serious risks—such as the potential for anyone to access protocols for designing bioweapons. Effective mitigation is unlikely to come from regulating AI itself; rather, authorities will need to lean on traditional frameworks, penalizing real-world misuse and controlling tangible precursor materials.

This entails some risks – like someone being able to easily get a protocol for designing a bioweapon. Effective mitigation is unlikely to come from regulating AI itself, due to difficulties of controlling its private use. Rather, authorities will need to lean on the traditional frameworks of imposing sanctions for and reducing opportunities for the tangible physical-world misconduct. For example, this could mean controlling instrumentation and precursor materials to mitigate chemical and biological threats.

7.3.5 Other predictions

Identity. AI's ability to impersonate humans at scale presents unprecedented risks of fraud and challenges for identity verification [CM24]. Robust authentication measures will become increasingly vital.

Advertising. There are two big trends that will deeply affect the advertising ecosystem. First, consumers have already started to use LLMs as a substitute for classical online search clicks [Kaf25; CL25], which will greatly affect the distribution of monetary flows in the advertising industry. There is already active research on how a brand could manipulate a third-party LLM into promoting the brand's content to the LLM users [NDT24]. Second, AI bots will continue to become more prevalent as both consumers and generators of online content [Hoo23]. In fact, we already live in the world where bots represent the majority of all the internet traffic [Tha25]. This means verified human attention will only grow in value – as will closed platform where users reveal their identities.

Autonomous weapons. Mosquito-sized, AI-controlled drones capable of delivering targeted toxins or explosives based on facial recognition are within reach [Man24]. AI techniques suited for control of swarms of drones have already been developed [GEK17]. Even unmanned fighter jets are already a reality [Cro25]. Countermeasure development must be an urgent priority – especially given the demonstrated use of drone warfare in conflicts like the Russia-Ukraine war, where up to 80% of the recent casualties have been caused by drone strikes [The25]. Unfortunately, but perhaps not unexpectedly, efforts to regulate "killer robots" at the United Nations have so far not yielded a result [Far21].

Agents' rights. Currently, most LLM models are wrapped in legal entities that deploy them. In a way, through willing human conduits, one could argue this might be effectively endowing LLMs with an ability to own assets and spend resources – as they become the "brains" of their organizations. Could we end up in a world, where LLMs could be granted basic rights even without humans participating – completely on their own? Perhaps – it will be very interesting to see how the law on this develops [SG25].

Data leaks. When one talks to an LLM that is not under one's full control, one should not assume the conversation will remain private. There is always a chance the conversation might leak – accidentally or by court order, for the public to read or for another LLM to train on. We have yet to witness the full scale of scandals and legal changes that may occur in this space. Perhaps, one day, the law will establish full AI-person confidentiality – akin to the privileged conversations one has with lawyers or doctors.

Heavy lobbying against AI. As AI disrupts established industries, vested interests will likely mount significant PR campaigns to slow or obstruct its adoption. Alarmism about AI risks is already quite popular [YS25; BSA25] and forms a fertile grounds for anti-AI campaigns. Large existing AI companies may take advantage of that sentiment, co-opting the governments to pursue more onerous AI regulation, which only large companies can afford to comply with, setting up a competitive moat against small entrants.

New form of life. It may be useful to regard advanced AI as a new, silicon-based life form, subject to evolutionary pressures and natural selection. Are humans just a stepping stone to this new form of life? Time will show.

New electricity. Even if the ambitions of the general AI are not realized, the "boring" effects of AI are already here – human productivity – from software engineers [Cui+24] to biological researchers [Qu+25] – has increased, vast data center infrastructure is being built out [Cas24], and the black box LLMs are churning out volumes of text, helping drive gradual automation of more and more human activities. A bit like electricity [Ng18], which we mostly no longer view as a miracle, even if LLMs fail to deliver regular "wow" moments, they are certainly here to stay as another staple of everyday human life.

Future of programming. In principle, once AI becomes sufficiently advanced and can execute arbitrary code on a server, it will be able to take even a vague product description and transform it into fully functional software and deploy it for users to enjoy. At that point, English itself would effectively become a programming language.

In practice, however, we are not there yet. Current LLMs still struggle with large, multi-file codebases [Voe25], though code editor providers such as Cursor are actively developing solutions to enable smoother workflows in these contexts. Moreover, deploying code as a working application exposed to real users involves substantial logistical effort – configuring credentials and payments, fine-tuning server settings, scaling infrastructure, and more. These tasks fall within the realm of DevOps (development operations), and as of this writing, no LLM has fully replaced a DevOps engineer. In fact, DevOps professionals are likely to remain useful longer than most other software specialists – after all, someone must ensure that the LLMs themselves continue to run smoothly.

At the same time, without a question, LLMs have already increased the productivity and transformed the workflow of software engineers [Cui+24]; a term *vibe coding* [Wikc] has been coined to describe the style of coding reliant on LLMs that involves trusting the AI to build what is needed, with minimal direct oversight – at least, at the initial stages. This transformation enables highly experienced engineers to operate with less reliance on junior support, leading to reduced demand for new software engineering hires, even as opportunities expand for solo developers and small teams to tackle big projects without large-scale human resource investment. As a result, the programming profession is certainly changing, but the opportunities for lucrative software engineer labor remain – at least, for now.

7.4 Discussion

Reflecting on the definitions at the start of this chapter, by many measures, AI already surpasses the average person in intelligence. For centuries, human cognition – limited by our brains' processing speed, the high cost of expertise, and uneven access to knowledge – has constrained progress in fields like scientific research, engineering, medicine, and the creative arts. With that bottleneck dissolving, the economic and institutional implications will be monumental.

Yet we need not view the impending automation and labor displacement as a catastrophe. Thoughtfully deployed, AI could vastly expand access to healthcare, education, legal counsel, and trustworthy information. It can liberate us from mundane tasks, freeing time for creativity and human connection.

Can we seize this opportunity wisely?

7.5 Further learning resources

- Science fiction:
    - Frank Herbert's *Dune* [Her65].
    - Harlan Ellison's *I Have No Mouth and I Must Scream* [Ell67].
    - Isaac Asimov's *I, Robot* [Asi50].
    - Robert Heinlein's *The Moon Is a Harsh Mistress* [Hei66].
    - William Gibson's *Neuromancer* [Gib84].
    - Jonathan Nolan and Lisa Joy's *Westworld (TV series)* [NJ16].
    - And many others.

- Demis Hassabis, 2024 Nobel Prize lecture in Chemistry, for accelerating scientific discovery with AI: https://www.nobelprize.org/prizes/chemistry/2024/hassabis/lecture/.

    - Review of AlphaFold 3 [Abr+24]: https://www.youtube.com/watch?v=qjFgthkKxcA.

- Geoffrey Hinton, 2024 Nobel Prize lecture in Physics, for Boltzmann machines (predecessors of modern neural nets): https://www.nobelprize.org/prizes/physics/2024/hinton/lecture/.

7.6 References

[Abr+24]   Josh Abramson et al. "Accurate structure prediction of biomolecular interactions with AlphaFold 3". In: *Nature* 630.8016 (2024), pp. 493–500.

[Ach+23]   Josh Achiam et al. "GPT-4 technical report". In: *arXiv preprint arXiv:2303.08774* (2023).

[AAV22]   Oscar Amador, Maytheewat Aramrattana, and Alexey Vinel. "A survey on remote operation of road vehicles". In: *IEEE Access* 10 (2022), pp. 130135–130154.

[Asc24]   Leopold Aschenbrenner. *Situational Awareness: The Decade Ahead*. 2024. URL: https://situational-awareness.ai/ (visited on 07/17/2025).

[Asi50]   Isaac Asimov. *I, Robot*. Gnome Press, 1950.

[Aut+24]   David Autor et al. "New frontiers: The origins and content of new work, 1940–2018". In: *The Quarterly Journal of Economics* 139.3 (2024), pp. 1399–1465.

[Aze+09]   Frederico AC Azevedo et al. "Equal numbers of neuronal and nonneuronal cells make the human brain an isometrically scaled-up primate brain". In: *Journal of Comparative Neurology* 513.5 (2009), pp. 532–541.

[Bai+22]   Yuntao Bai et al. "Training a helpful and harmless assistant with reinforcement learning from human feedback". In: *arXiv preprint arXiv:2204.05862* (2022).

[Bar+24]   Omer Bar-Tal et al. "Lumiere: A space-time diffusion model for video generation". In: *SIGGRAPH Asia 2024*. 2024, pp. 1–11.

[BSA25]   Peter Barnett, Aaron Scher, and David Abecassis. *Technical Requirements for Halting Dangerous AI Activities*. 2025. arXiv: 2507.09801. URL: https://arxiv.org/abs/2507.09801.

[Bid19]   Juliana Uhuru Bidadanure. "The political theory of universal basic income". In: *Annual Review of Political Science* 22.1 (2019), pp. 481–501.

[Bil24]   Massimo Bilancioni. *Energy efficiency: AI vs. the human brain.* 2024. URL: https://medium.com/@massimo.bilancioni21/ener gy-efficiency-ai-vs-human-brain-8a6fc5488492#478e.

[Bri+25]   Garyk Brixi et al. "Genome modeling and design across all domains of life with Evo 2". In: *BioRxiv* (2025), pp. 2025–02.

[Bro+23]   Anthony Brohan et al. "RT-2: Vision-language-action models transfer web knowledge to robotic control". In: *arXiv preprint arXiv:2307.15818* (2023).

[Cas24]   Daniel Castro. *Rethinking Concerns About AI's Energy Use.* 2024. URL: https://www2.datainnovation.org/2024-ai-energy-use.pdf.

[Cha97]   David J Chalmers. *The conscious mind: In search of a funda-mental theory.* Oxford Paperbacks, 1997.

[CL25]   Athena Chapekis and Anna Lieb. *Google users are less likely to click on links when an AI summary appears in the results.* 2025. URL: https://www.pewresearch.org/short-reads/2025/0 7/22/google-users-are-less-likely-to-click-on-links-when-an-ai-s ummary-appears-in-the-results/ (visited on 07/23/2025).

[CM24]   Heather Chen and Kathleen Magramo. "Finance worker pays out $25 million after video call with deepfake 'chief financial officer'". In: *CNN* (2024). Accessed 2025-07-07. URL: https: //www.cnn.com/2024/02/04/asia/deepfake-cfo-scam-hong-ko ng-intl-hnk.

[CV25]   Lihu Chen and Gaël Varoquaux. *What is the Role of Small Models in the LLM Era: A Survey.* 2025. arXiv: 2409.06857. URL: https://arxiv.org/abs/2409.06857.

[Chi+25]   Wayne Chi et al. "Copilot Arena: A platform for code LLM evaluation in the wild". In: *arXiv preprint arXiv:2502.09328* (2025).

[Chi+24]   Wei-Lin Chiang et al. "Chatbot Arena: An open platform for evaluating LLMs by human preference". In: *ICML.* 2024.

[Cho+24]   Neil Chowdhury et al. *Introducing SWE-bench Verified.* Ope-nAI Blog. 2024. URL: https://openai.com/index/introducing-s we-bench-verified/ (visited on 07/26/2025).

[Cla13]   Andy Clark. "Whatever next? Predictive brains, situated agents, and the future of cognitive science". In: *Behavioral and Brain Sciences* 36.3 (2013), pp. 181–204.

[Cla25]   Jack Clark. *Import AI 421: Kimi 2 – a great Chinese open weight model; giving AI systems rights and what it means; and how to pause AI progress*. 2025. URL: https://jack-clark.net/2025/07/21/import-ai-421-kimi-2-a-great-chinese-open-weight-model-giving-ai-systems-rights-and-what-it-means-and-how-to-pause-ai-progress/ (visited on 07/23/2025).

[CAS16]   Paul Covington, Jay Adams, and Emre Sargin. "Deep neural networks for YouTube recommendations". In: *ACM RecSys*. 2016, pp. 191–198.

[Cro25]   Will Croxton. *Anduril CEO unveils the Fury unmanned fighter jet*. 2025. URL: https://www.cbsnews.com/news/anduril-ceo-unveils-the-fury-unmanned-fighter-jet-60-minutes/ (visited on 07/17/2025).

[Cui+24]   Kevin Zheyuan Cui et al. "The Productivity Effects of Generative AI: Evidence from a Field Experiment with GitHub Copilot". In: (2024).

[De 23]   Alex De Vries. "The growing energy footprint of artificial intelligence". In: *Joule* 7.10 (2023), pp. 2191–2194.

[Di +24]   Luigi Di Lillo et al. *Do autonomous vehicles outperform latest-generation human-driven vehicles? A comparison to Waymo's auto liability insurance claims at 25.3 M miles*. 2024.

[Dra05]   David A Drachman. *Do we have brain to spare?* 2005.

[Ell67]   Harlan Ellison. *I have no mouth & I must scream*. Galaxy Publishing Corp, 1967.

[Far21]   Emma Farge. "U.N. talks adjourn without deal to regulate 'killer robots'". In: (2021). URL: https://www.reuters.com/world/un-talks-adjourn-without-deal-regulate-killer-robots-2021-12-17/ (visited on 07/31/2025).

[Faw+22]   Alhussein Fawzi et al. "Discovering faster matrix multiplication algorithms with reinforcement learning". In: *Nature* 610.7930 (2022), pp. 47–53.

[Fox24]   Jacob Fox. *You can now rent a 'minibrain' for $500 a month and the potential implications are horrifying.* 2024. URL: https://www.pcgamer.com/hardware/you-can-now-rent-a-minibrain-for-dollar500-a-month-and-the-potential-implications-are-horrifying/ (visited on 07/20/2025).

[Fri10]   Karl Friston. "The free-energy principle: A unified brain theory?" In: *Nature Reviews Neuroscience* 11.2 (2010), pp. 127–138.

[Gen+22]   Shijie Geng et al. "Recommendation as language processing (RLP): A unified pretrain, personalized prompt & predict paradigm (P5)". In: *ACM RecSys.* 2022, pp. 299–315.

[Gib84]   William Gibson. *Neuromancer.* Ace Books, 1984.

[Gil+17]   Jill Gilkerson et al. "Mapping the early language environment using all-day recordings and automated analysis". In: *American Journal of Speech-Language Pathology* 26.2 (2017), pp. 248–265.

[Gla+24]   Elliot Glazer et al. "Frontiermath: A benchmark for evaluating advanced mathematical reasoning in ai". In: *arXiv preprint arXiv:2411.04872* (2024).

[Gra+24]   Aaron Grattafiori et al. "The Llama 3 herd of models". In: *arXiv preprint arXiv:2407.21783* (2024).

[Guo+24]   Taicheng Guo et al. "Large language model based multi-agents: A survey of progress and challenges". In: *arXiv preprint arXiv:2402.01680* (2024).

[GEK17]   Jayesh K Gupta, Maxim Egorov, and Mykel Kochenderfer. "Cooperative multi-agent control using deep reinforcement learning". In: *AAMAS.* 2017, pp. 66–83.

[HK19]   Michael Haenlein and Andreas Kaplan. "A brief history of artificial intelligence: On the past, present, and future of artificial intelligence". In: *California Management Review* 61.4 (2019), pp. 5–14.

[Han+25]   David Hanson et al. *Sentience Quest: Towards Embodied, Emotionally Adaptive, Self-Evolving, Ethically Aligned Artificial General Intelligence.* 2025. arXiv: 2505.12229. URL: https://arxiv.org/abs/2505.12229.

[Hei66]   Robert A Heinlein. *The moon is a harsh mistress*. G. P. Putnam's Sons, 1966.

[Hen+20]   Dan Hendrycks et al. "Measuring massive multitask language understanding". In: *arXiv preprint arXiv:2009.03300* (2020).

[Her65]   Frank Herbert. "Dune". In: Chilton Books, 1965.

[Hoo23]   Amanda Hoover. *Spotify Has an AI Music Problem – but Bots Love It*. 2023. URL: https://www.wired.com/story/spotify-ai-music-robot-listeners/ (visited on 07/20/2025).

[Hwa+24]   Jyh-Jing Hwang et al. "EMMA: End-to-end multimodal model for autonomous driving". In: *arXiv preprint arXiv:2410.23262* (2024).

[IMO25]   IMO. *66th IMO 2025*. 2025. URL: https://www.imo-official.org/year_info.aspx?year=2025 (visited on 07/23/2025).

[Jim+23]   Carlos E Jimenez et al. "SWE-bench: Can language models resolve real-world GitHub issues?" In: *arXiv preprint arXiv:2310.06770* (2023).

[Kaf25]   Peter Kafka. *Apple says searches are shrinking because people are using AI instead. Now Google's stock is tanking*. 2025. URL: https://www.businessinsider.com/apple-says-ai-disrupts-search-market-google-impact-2025-5 (visited on 07/20/2025).

[Kag+22]   Brett J Kagan et al. "In vitro neurons learn and exhibit sentience when embodied in a simulated game-world". In: *Neuron* 110.23 (2022), pp. 3952–3969.

[Kam+19]   Juliane Kaminski et al. "Evolution of facial muscle anatomy in dogs". In: *PNAS* 116.29 (2019), pp. 14677–14681.

[Kap+20]   Jared Kaplan et al. "Scaling laws for neural language models". In: *arXiv preprint arXiv:2001.08361* (2020).

[Khr+24]   Qusai Khraisha et al. "Can large language models replace humans in systematic reviews? Evaluating GPT-4's efficacy in screening and extracting data from peer-reviewed and grey literature in multiple languages". In: *Research Synthesis Methods* 15.4 (2024), pp. 616–626.

[Kic+23]   Emre Kiciman et al. "Causal reasoning and large language models: Opening a new frontier for causality". In: *Transactions on Machine Learning Research* (2023).

[Kim25]   Kimi Team. *Kimi K2: Open Agentic Intelligence*. 2025. URL: https://github.com/MoonshotAI/Kimi-K2 (visited on 07/23/2025).

[Kus+25]   Kristofer D Kusano et al. "Comparison of Waymo Rider-Only crash rates by crash type to human benchmarks at 56.7 million miles". In: *Traffic Injury Prevention* (2025), pp. 1–13.

[Kut+25]   Jonathan Kutasov et al. "SHADE-Arena: Evaluating Sabotage and Monitoring in LLM Agents". In: *arXiv preprint arXiv:2506.15740* (2025).

[Liu+24]   Aixin Liu et al. "DeepSeek-V3 technical report". In: *arXiv preprint arXiv:2412.19437* (2024).

[Liu+25]   Yixiu Liu et al. "AlphaGo Moment for Model Architecture Discovery". In: *arXiv preprint arXiv:2507.18074* (2025).

[Liu+22]   Zhuoran Liu et al. "Monolith: Real time recommendation system with collisionless embedding table". In: *arXiv preprint arXiv:2209.07663* (2022).

[Lu+25]   Sining Lu et al. *LL3M: Large Language 3D Modelers*. 2025. arXiv: 2508.08228. URL: https://arxiv.org/abs/2508.08228.

[Luo+25]   Xiaoliang Luo et al. "Large language models surpass human experts in predicting neuroscience results". In: *Nature Human Behaviour* 9.2 (2025), pp. 305–315.

[LL25]   Thang Luong and Edward Lockhart. *Advanced version of Gemini with Deep Think officially achieves gold-medal standard at the International Mathematical Olympiad*. 2025. URL: https://deepmind.google/discover/blog/advanced-version-of-gemini-with-deep-think-officially-achieves-gold-medal-standard-at-the-international-mathematical-olympiad/ (visited on 07/23/2025).

[Man24]   Nora Mankel. "Chinese military robotics lab creates mosquito-sized microdrone for covert operations". In: *South China Morning Post* (2024). Accessed 2025-07-07. URL: https://www.scmp.com/news/china/science/article/3315206/chinese-military-robotics-lab-creates-mosquito-sized-microdrone-covert-operations.

[McC+06]   John McCarthy et al. "A proposal for the Dartmouth summer research project on artificial intelligence, August 31, 1955". In: *AI magazine* 27.4 (2006), pp. 12–12.

[MC04]   Pamela McCorduck and Cli Cfe. *Machines Who Think: A Personal Inquiry Into the History and Prospects of Artificial Intelligence.* AK Peters/CRC Press, 2004.

[McD+25]   Daniel McDuff et al. "Towards accurate differential diagnosis with large language models". In: *Nature* (2025), pp. 1–7.

[Mil56]   George A Miller. "The magical number seven, plus or minus two: Some limits on our capacity for processing information." In: *Psychological review* 63.2 (1956), p. 81.

[Mis+24]   Anand Kumar Mishra et al. "Sensorimotor control of robots mediated by electrophysiological measurements of fungal mycelia". In: *Science Robotics* 9.93 (2024), eadk8019.

[Naj23]   Reabal Najjar. "Redefining radiology: A review of artificial intelligence integration in medical imaging". In: *Diagnostics* 13.17 (2023), p. 2760.

[Nel+98]   Robert C Nelson et al. *Flight Stability and Automatic Control.* WCB / McGraw Hill New York, 1998.

[NDT24]   Fredrik Nestaas, Edoardo Debenedetti, and Florian Tramèr. "Adversarial search engine optimization for large language models". In: *arXiv preprint arXiv:2406.18382* (2024).

[Ng18]   Andrew Ng. *AI is the new electricity.* O'Reilly Media, 2018.

[NJ16]   Jonathan Nolan and Lisa Joy. *Westworld.* HBO, 2016–2022.

[Ope24]   OpenAI. *GPT-4o mini: Advancing cost-efficient intelligence.* 2024. URL: https://openai.com/index/gpt-4o-mini-advancing-cost-efficient-intelligence/.

[Ope25]   OpenAI. *OpenAI o3 and o4-mini System Card*. Tech. rep. OpenAI, 2025.

[OL25]   C Opus and A Lawsen. "The Illusion of the Illusion of Thinking". In: *arXiv preprint ArXiv:2506.09250* (2025).

[Pri23]   Jacopo Prisco. *This cargo plane flew with no pilot on board*. 2023. URL: https://www.cnn.com/travel/cessna-cargo-plane-flight-no-pilot-on-board-spc-intl (visited on 07/17/2025).

[Qu+25]   Yuanhao Qu et al. "CRISPR-GPT for agentic automation of gene-editing experiments". In: *Nature Biomedical Engineering* (2025), pp. 1–14.

[Rei+24]   David Rein et al. "GPQA: A graduate-level Google-proof Q&A benchmark". In: *COLM*. 2024.

[Roo24]   Kevin Roose. "A.I. Has a Measurement Problem". In: *The New York Times* (Apr. 15, 2024). Accessed 2025-07-07. URL: https://www.nytimes.com/2024/04/15/technology/ai-models-measurement.html.

[SG25]   Peter Salib and Simon Goldstein. "AI Rights for Human Flourishing". SSRN. 2025. DOI: 10.2139/ssrn.5353214. URL: https://ssrn.com/abstract=5353214.

[Sch23]   Maximilian Schreiner. *GPT-4 architecture, datasets, costs and more leaked*. 2023. URL: https://the-decoder.com/gpt-4-architecture-datasets-costs-and-more-leaked/ (visited on 05/30/2025).

[Ser+20]   Pedro Jesus Serrano-Castro et al. "Neuroplasticity and epilepsy surgery in brain eloquent areas: Case report". In: *Frontiers in Neurology* 11 (2020), p. 549172.

[SHF23]   Mohammad Amin Shabani, Sepidehsadat Hosseini, and Yasutaka Furukawa. "HouseDiffusion: Vector floorplan generation via a diffusion model with discrete and continuous denoising". In: *IEEE/CVF CVPR*. 2023, pp. 5466–5475.

[Sha24]   Max Shap. *How Long Does It Take to Train the LLM From Scratch?* 2024. URL: https://towardsdatascience.com/how-long-does-it-take-to-train-the-llm-from-scratch-a1adb194c624/.

[She+25] Tao Shen et al. "Will LLMs scaling hit the wall? Breaking barriers via distributed resources on massive edge devices". In: *arXiv preprint arXiv:2503.08223* (2025).

[Sho+25] Parshin Shojaee et al. "The illusion of thinking: Understanding the strengths and limitations of reasoning models via the lens of problem complexity". In: *arXiv preprint arXiv:2506.06941* (2025).

[Sil+16] David Silver et al. "Mastering the game of Go with deep neural networks and tree search". In: *Nature* 529.7587 (2016), pp. 484–489.

[Sil+18] David Silver et al. "A general reinforcement learning algorithm that masters chess, shogi, and Go through self-play". In: *Science* 362.6419 (2018), pp. 1140–1144.

[Sun+19] Fei Sun et al. "BERT4Rec: Sequential recommendation with bidirectional encoder representations from transformer". In: *ACM CIKM.* 2019, pp. 1441–1450.

[Tha25] Thales Group. *Bad Bots on the Rise: Internet Traffic Hits Record Levels.* 2025. URL: https://www.thalesgroup.com/en/w orldwide/digital-identity-and-security/magazine/bad-bots-rise -internet-traffic-hits-record-levels (visited on 08/15/2025).

[The24] The Associated Press. *OpenAI appoints former top US cyber-warrior Paul Nakasone to its board of directors.* 2024. URL: https://apnews.com/article/openai-nsa-director-paul-nakason e-cyber-command-6ef612a3a0fcaef05480bbd1ebbd79b1 (visited on 07/20/2025).

[The25] The Week US. "Death from Above: Drones Upend Rules of War in Ukraine". In: (2025). URL: https://theweek.com/p olitics/death-drones-upend-rules-war-ukraine (visited on 07/31/2025).

[Tka24] Yegor Tkachenko. "Position: Enforced Amnesia as a Way to Mitigate the Potential Risk of Silent Suffering in the Conscious AI". In: *ICML.* 2024.

[Tom+25] Kiran Tomlinson et al. *Working with AI: Measuring the Occupational Implications of Generative AI.* 2025. arXiv: 2507.079 35. URL: https://arxiv.org/abs/2507.07935.

[Tou+23] Hugo Touvron et al. "Llama: Open and efficient foundation language models". In: *arXiv preprint arXiv:2302.13971* (2023).

[Vaf+25] Keyon Vafa et al. "What Has a Foundation Model Found? Using Inductive Bias to Probe for World Models". In: *arXiv preprint arXiv:2507.06952* (2025).

[Var25] Lakshmi Varanasi. *OpenAI just won gold at the world's most prestigious math competition. Here's why that's a big deal.* 2025. URL: https://www.businessinsider.com/openai-gold-iom-math-competition-2025-7 (visited on 07/23/2025).

[Var+25] Iñaki Dellibarda Varela et al. "Rethinking the Illusion of Thinking". In: *arXiv preprint arXiv:2507.01231* (2025).

[Vas+17] Ashish Vaswani et al. "Attention is All You Need". In: *NeurIPS*. 2017. URL: https://arxiv.org/pdf/1706.03762.pdf.

[Voe25] Colton Voege. *No, AI is not Making Engineers 10x as Productive: Curing Your AI 10x Engineer Imposter Syndrome.* 2025. URL: https://colton.dev/blog/curing-your-ai-10x-engineer-imposter-syndrome/ (visited on 08/16/2025).

[Wan+24a] Jiaqi Wang et al. "A comprehensive review of multimodal large language models: Performance and challenges across different tasks". In: *arXiv preprint arXiv:2408.01319* (2024).

[Wan+23] Weizhi Wang et al. "Augmenting language models with long-term memory". In: *NeurIPS* 36 (2023), pp. 74530–74543.

[Wan+24b] Yubo Wang et al. "MMLU-Pro: A more robust and challenging multi-task language understanding benchmark". In: *NeurIPS*. 2024.

[War+23] Alex Warstadt et al. "Findings of the BabyLM Challenge: Sample-Efficient Pretraining on Developmentally Plausible Corpora". In: *CoNLL*. Association for Computational Linguistics, 2023. DOI: 10.18653/v1/2023.conll-babylm.1. URL: http://dx.doi.org/10.18653/v1/2023.conll-babylm.1.

[Wei+22] Jason Wei et al. "Chain-of-thought prompting elicits reasoning in large language models". In: *NeurIPS* 35 (2022), pp. 24824–24837.

[Wika] Wikipedia. *Goodhart's law.* URL: https://en.wikipedia.org/wiki/Goodhart%27s_law (visited on 07/20/2025).

[Wikb]   Wikipedia. *SVG*. URL: https://en.wikipedia.org/wiki/SVG (visited on 07/23/2025).

[Wikc]   Wikipedia. *Vibe coding*. URL: https://en.wikipedia.org/wiki /Vibe_coding (visited on 08/17/2025).

[Wil25]   Simon Willison. *Import AI 421: Kimi 2 – a great Chinese open weight model; giving AI systems rights and what it means; and how to pause AI progress*. 2025. URL: https://simonwillison.ne t/2025/Jul/11/kimi-k2/ (visited on 07/23/2025).

[Xie+25]   Yichen Xie et al. "S4-Driver: Scalable Self-Supervised Driving Multimodal Large Language Model with Spatio-Temporal Visual Representation". In: *IEEE/CVF CVPR*. 2025, pp. 1622–1632.

[YS25]   Eliezer Yudkowsky and Nate Soares. *If Anyone Builds It, Everyone Dies: Why Superhuman AI Would Kill Us All*. New York, NY: Little, Brown and Company, 2025. ISBN: 9780316595643.

[Zhe+23]   Lianmin Zheng et al. "Judging LLM-as-a-judge with MT-Bench and Chatbot Arena". In: *NeurIPS* 36 (2023), pp. 46595–46623.

About the author

Kauai, Hawaii, 2024

Yegor Tkachenko is an Adjunct Assistant Professor at Columbia Business School in the New York City, where he created the *Python Programming for Data Science* course that inspired this book. He holds a PhD from Columbia University and an MS from Stanford University.

Personal website: https://yegortkachenko.com/.

Index

www.ingramcontent.com/pod-product-compliance
Lightning Source LLC
Chambersburg PA
CBHW060745220326
41598CB00022B/2328